The ESSENTIALS® of

REGISTERED TRADEMARK

MOLECULAR STRUCTURES OF LIFE

Maya Pines and
Anne A. Oplinger
U.S. National Institute
of General Medical Sciences

REA

Research and Education Association
61 Ethel Road West
Piscataway, New Jersey 08854

THE ESSENTIALS® OF
MOLECULAR STRUCTURES OF LIFE

Printed in the United States of America

Library of Congress Catalog Card Number 95-67789

International Standard Book Number 0-87891-983-X

I-1

WHAT "THE ESSENTIALS" WILL DO FOR YOU

This book is a review and study guide. It is comprehensive and it is concise.

It helps in preparing for exams, in doing homework, and remains a handy reference source at all times.

It condenses the vast amount of detail characteristic of the subject matter and summarizes the **essentials** of the field.

It will thus save hours of study and preparation time.

The book provides quick access to the important principles, concepts, theories and practices in the field.

Materials needed for exams can be reviewed in summary form – eliminating the need to read and re-read many pages of textbook and class notes. The summaries will even tend to bring detail to mind that had been previously read or noted.

This "ESSENTIALS" book has been prepared by an expert in the field, and has been carefully reviewed to assure accuracy and maximum usefulness.

Dr. Max Fogiel
Program Director

CONTENTS

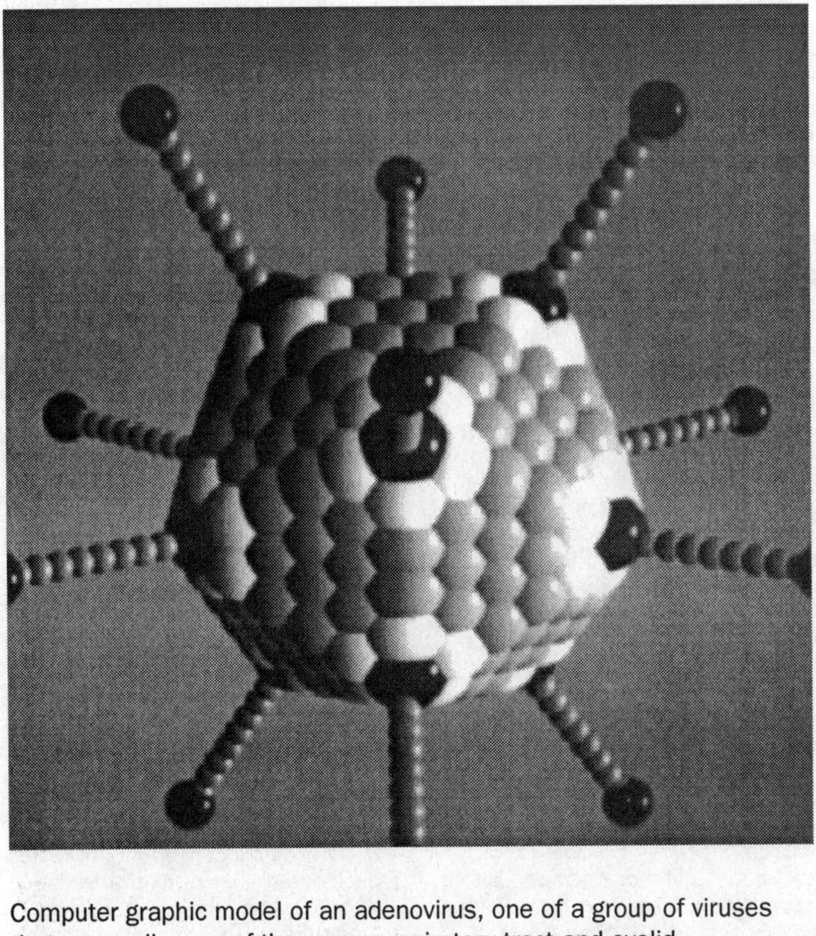

Computer graphic model of an adenovirus, one of a group of viruses that cause disease of the upper respiratory tract and eyelid.

THE PROMISE AND EXCITEMENT OF STRUCTURAL BIOLOGY

A scientist in Indiana has found a vulnerable spot in one of the viruses that cause the common cold—and as a result, a real cure for colds may now be possible. Other scientists are working around the clock to find a similar chink in the armor of the AIDS virus.

Some researchers are laying the groundwork for a new class of safer, more effective cancer drugs. And another group of researchers is on its way to learning how to switch on specific genes when they are needed by the body—as, for example, to regenerate damaged tissues after an injury.

What these investigators have in common is their focus on fundamental biological shapes. Their work is at the forefront of a rapidly growing science called structural biology, which aims to explain the activity of biologically important molecules in terms of their atomic structure—and to use this knowledge to design new therapies or vaccines.

When scientists understand the precise structure of some components of the surface of cancer cells, for instance, they may be able to tailor new drugs to fit these components as accurately as a key in a lock. With a really tight fit, very low doses of drugs could kill the cancer cells effectively without harming the normal cells nearby.

Similarly, researchers are trying to understand the three-dimensional (3-D) structure of some natural chemicals that attach themselves to the genetic material DNA. These chemicals selectively control the activity of specific genes at specific times and in specific places. Deciphering their structure could lead to laboratory-made chemicals that do exactly the same thing.

Structural biologists use information gathered from chemistry, physics, genetics, cell biology, and mathematics, plus supercomputers and other highly specialized equipment. In recent years these scientists have relied increasingly on the use of computer graphics to visualize how various subunits of molecules fit together or move.

They have made rapid strides in the analysis of proteins (the extraordinarily varied molecules that do most of the work of the body); nucleic acids (DNA and a related molecule, RNA); carbohydrates (sugars and starches); lipids (fats); and the complex combinations of these substances.

Structural biology burst upon public attention for the first time in 1953, when Francis H.C. Crick and James D. Watson announced that they had deciphered the structure of DNA—work for which they later won the Nobel Prize. Their model, the famous double helix, paved the way for many advances in genetics, including the development of recombinant DNA technology, which allows scientists to cut and splice together pieces of DNA from different sources.

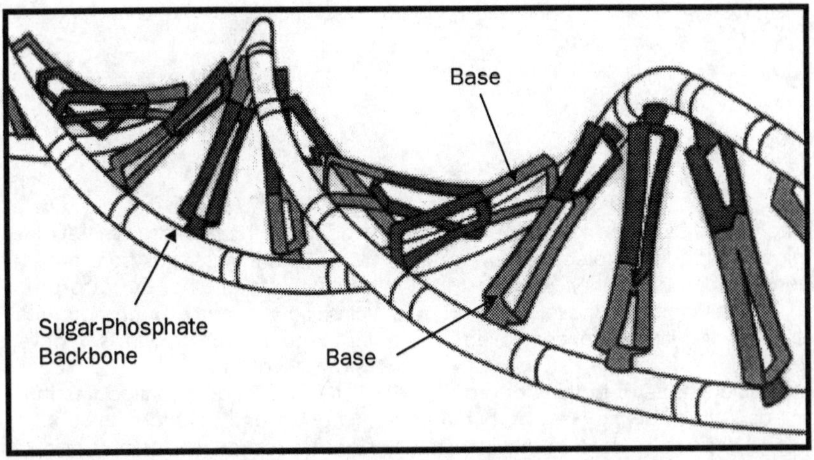

The double helix: model of the DNA molecule that embodies the code of heredity, as conceived by Francis H.C. Crick and James D. Watson in 1953. All living things reproduce themselves according to the genetic information in different sequences of DNA subunits (bases).

According to this model, DNA is a twisting ladder made up of different sequences of four components called nucleotides. Each rung of the ladder consists of a pair of nucleotides (which can pair only in certain ways). The specific sequence of these pairs contains all the information necessary for the development and survival of an organism such as a human being. The model explained for the first time how genetic information is transferred from the parent archive of DNA to daughter strands of nucleic acid: the nucleotide pairs separate in the middle of the rung, as if a zipper had opened, making two complementary strands, and each of these strands participates in forming a new, complete double helix.

The structures of many other biologically important molecules have been revealed since then, particularly the structures of those marvelously versatile substances called proteins.

Nonscientists tend to think of proteins as just something we eat. But in fact we depend on tens of thousands of different proteins in our bodies to keep us alive. Each protein is a complex biochemical machine with its own specialty. The reason we eat proteins from plants or animals is that we need the raw materials to make proteins of our own.

Some of the proteins we depend on are enzymes that increase the speed of chemical reactions up to a million times, without themselves undergoing any change. In this way, enzymes control the pathways and the timing of billions of chemical operations. They regulate our growth from a single cell to a mature organism. They make our cells differentiate into eye, blood, or brain cells. They transform our food into energy. They deftly break down or build up other proteins.

There are other proteins that sit on the surface membranes of our cells, where they receive messages from distant cells, or control the flow of molecules into or out of the cell. The surfaces of cancer cells are dotted with proteins that play a large role in the cells' runaway growth. The coats of viruses are made up largely of proteins.

Some of the neurotransmitters that carry urgent messages from one nerve cell to another are also proteins. So are many of the hormones that regulate our growth, sex drive, and reactions to stress.

The thousands of antibodies that recognize and fight foreign substances such as viruses and bacteria are proteins. The chemicals that interact with DNA to turn specific genes on or off are proteins. One protein, hemoglobin, has the key job of transporting oxygen throughout the body via the bloodstream.

Different kinds of proteins act as structural materials, making up our skin, hair, nails, muscles, tendons, and bones. They also maintain the inner structure of the millions of cells in our bodies.

Some proteins are relatively small and compact; others are bulky conglomerates. Some are globular; others are long and narrow. But all of these different types of proteins are polymers, substances composed of many smaller units linked in chains. The subunits of proteins are 20 different amino acids, which can be linked together in any order. Thus, for any protein made up of 100 amino acids there can be a nearly infinite number of possible structures. But each protein has a specific order, and one protein can be distinguished from another by the order of its amino acids.

Difficult as it was to "solve" the structure of DNA, protein structures

a

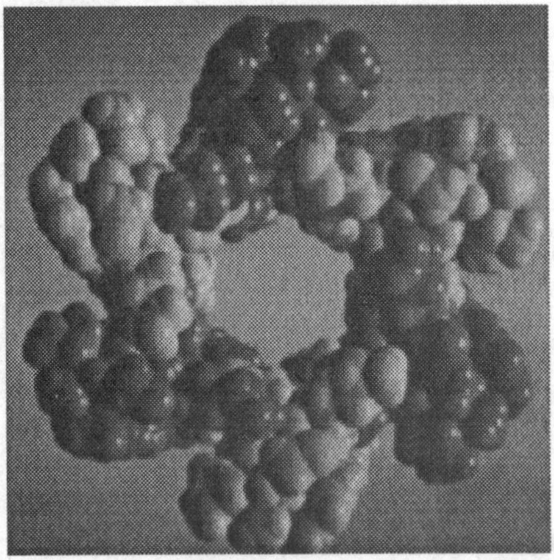

b

Proteins come in many shapes, according to their functions: bovine trypsin inhibitor, an enzyme inhibitor (a); a ring of six peptides of magainin, an antimicrobial substance found in frog skin (b); human immunoglobulin G, an antibody (c).

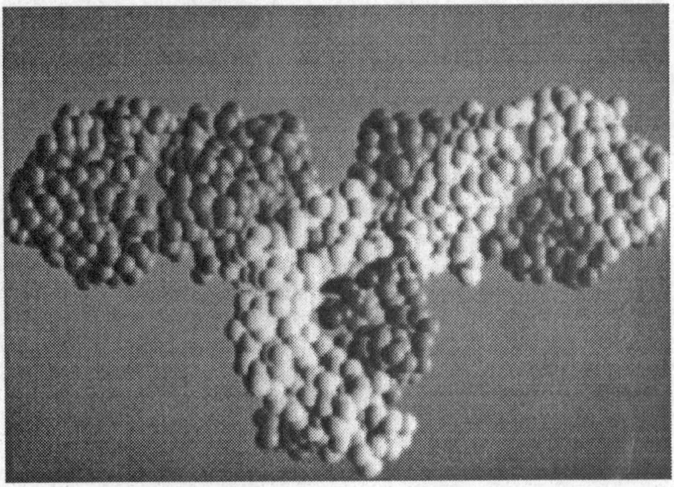

c

are far more difficult to decipher—not only because proteins have more subunits, but because they have irregular shapes. DNA structures tend to be more regular, although variations can still occur. By contrast, each type of protein molecule has a different shape, which determines its function. This shape defines what other molecules will bind to the protein and what chemical activity will take place.

All proteins are constructed in accordance with instructions coded by the DNA inside our cells. "Your genes supply the information," explains Russell Doolittle of the University of California, San Diego, "but you *are* your proteins." Nevertheless, we have only begun to understand how proteins function. The toughest problem has been to decipher their 3-D structure, which holds the key to their normal or abnormal activity.

Analyzing the structures of proteins and nucleic acids down to the precise arrangement of their atoms, and trying to understand the basis of each protein's activity, are major goals of structural biology. For a while these goals seemed very distant, but now the field is in a period of great excitement because of the new possibilities opened up by several recent developments:

- Recombinant DNA technology is allowing scientists to obtain large quantities of specific proteins or nucleic acids for study. In the past, research was severely limited by the shortage of experimental material.
- Researchers can now determine the sequence of subunits of these substances—for example, the sequence of amino acids in specific proteins, or the sequence of nucleotides in specific stretches of DNA—without an extraordinary expenditure of time and effort.
- Major improvements in X-ray crystallography, nuclear magnetic resonance (NMR) spectroscopy, and other techniques are enabling researchers to decipher the 3-D structures of large molecules more precisely and rapidly.
- Faster and less expensive computers are giving scientists new powers to handle enormous quantities of data and—through computer graphics—to visualize the structure and movement of proteins.

The coming together of these separate developments has made structural biology ready for a major leap forward. So promising is this work that many research laboratories in universities and drug companies are rushing into the field, and a shortage of specialists has developed.

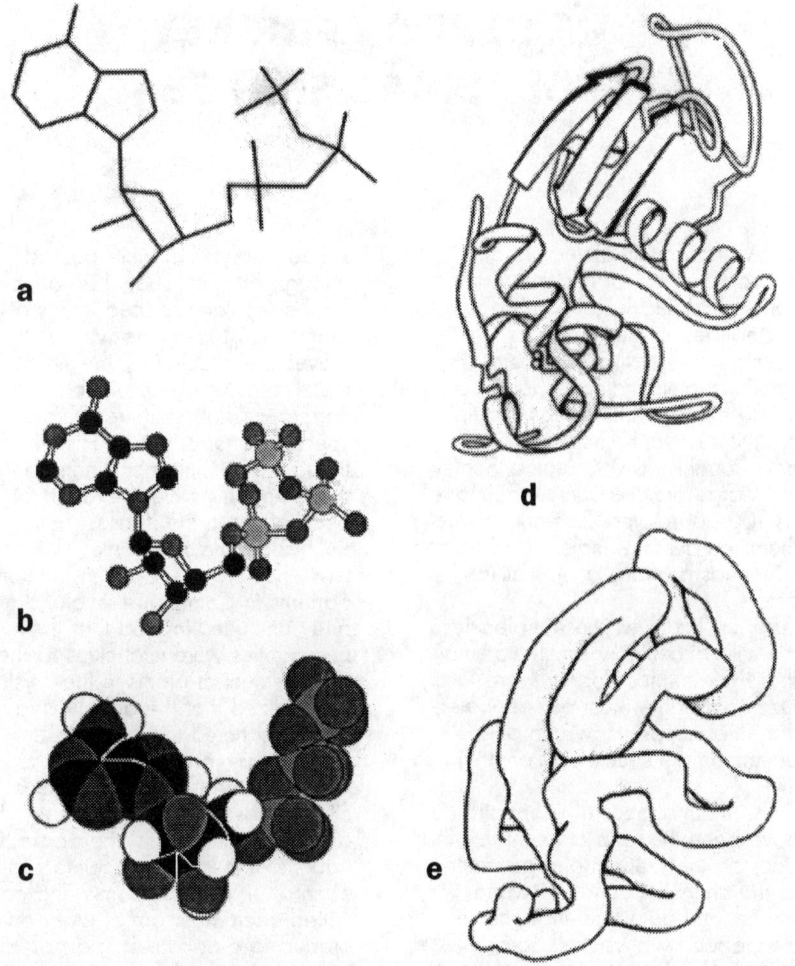

Molecular structures are represented by various types of models. The skeletal model (a), in which the positions of atoms are implied by the junctions and termini of bonds, is used to show molecular framework. The ball-and-stick model (b) represents bonds (by sticks) more clearly. The space-filling model (c), which represents the size and configuration of an atom as determined by its bonding properties, is the most realistic. The "ribbon model" (d) represents regions of regular hydrogen-bonded interactions as coils or sets of arrows. The "sausage model" (e) shows the course of the polypeptide chain but omits all other detail. Note that the core of a globular protein is densely packed with atoms. Thus, the impression of an open structure given by some models is misleading.

THE SEARCH FOR KEY BIOLOGICAL SHAPES

DNA and protein molecules are called ''macro,'' or large, in comparison with molecules of water, fats, nucleotides, or amino acids, which consist of fewer than 50 atoms. A smaller protein molecule has about 1,000 atoms, while a more complex protein such as hemoglobin may have roughly 10,000 atoms. Some very large proteins contain as many as 100,000 atoms. Towering above them, an average molecule of human DNA has as many as 4.4 billion atoms.

But in fact, even these molecules are still exceedingly small—so small that their images appear weak and fuzzy under the most powerful electron microscope. How can one determine the structure of anything so tiny?

The main sources of information have been the laws of chemistry that bear on the strength and geometry of the chemical bonds between atoms, and data from instruments developed by physicists. Some of the most useful data have come from X-ray crystallography, a technique based on the fact that the atoms in a crystal are arranged in a definite pattern that is repeated regularly in three dimensions.

As early as 1912, Max von Laue, a German physicist, realized that the wavelengths of X-rays—3,000 to 4,000 times shorter than those of visible light—were just about the same size as the spaces between atoms in a crystal. Therefore, he reasoned, X-rays that were passed through a crystal should be affected by the regularly spaced layers of atoms in just the same way as light waves are affected by the slats of a metal grid. Von Laue's idea proved correct: crystals scattered X-rays in specific patterns. But he mistakenly thought that some peculiarities in these patterns were the results of peculiarities in the X-rays.

At about the same time, William Lawrence Bragg looked at the same patterns in Cambridge, England, and concluded instead that their peculiarities were vital clues to the arrangement of atoms in the crystal. He realized that the orderly array of atoms scattered the X-rays in an orderly way, and that this caused a repeating series of overlapping circles of waves. When intercepted by a photographic plate, the peaks and troughs of these waves reinforced each other at some points and cancelled each other out at other points, producing a pattern of spots of varying intensity—an interference pattern that could be translated into information about the structure that produced it.

Bragg rapidly worked out equations with which scientists could get an image of the arrangement of atoms in a crystal from the X-rays diffracted by a particular axis of the crystal. If one keeps realigning the crystal so that a different axis of the crystal is parallel to the X-ray beam each time, one can make a set of

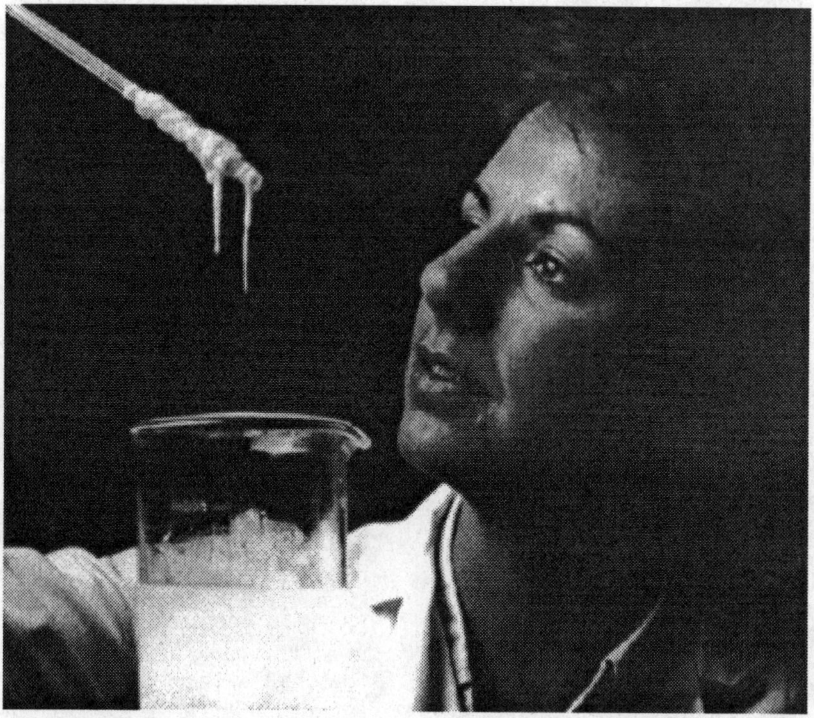

DNA, which has been extracted from cells of the bacterium *E. coli*, on the end of a glass rod.

photographs which, with the aid of Bragg's equations, allow one to deduce the 3-D arrangements of the molecules in the crystal.

Bragg soon tried out his technique on crystals of sodium chloride (table salt) and potassium chloride and succeeded in working out their atomic structures. In 1915, at the unusually young age of 25, he won the Nobel Prize for having invented X-ray crystallography and "solved" the first structures of crystals.

In order to do X-ray diffraction studies of a biological substance, one must get the molecules neatly aligned to form straight planes that can reflect the X-ray beams. This generally means that the substance must take the form of a crystal. But in some cases the fibers of a substance will naturally be so neatly aligned that they behave as though they were well-ordered crystals. Fortunately, this is what happened in some of the first experiments with X-ray diffraction of DNA.

DNA is an extremely long, thin, and fragile thread that scientists have had great difficulty in extracting from cell nuclei in pieces that are large enough to study. While working with a highly viscous solution of DNA extracted from calf thymus cells in 1950, Maurice Wilkins, a British biophysicist at King's College in London, noticed that every time he touched this solution with the tip of a glass rod and then drew the rod away, he "had spun a very thin fibre of DNA, almost invisible, like a filament of spider web." The fibers seemed highly uniform, so Wilkins took them to a graduate student for X-ray crystallography. These first X-ray diffraction pictures of DNA offered many clues to its structure.

James Watson, an American who had come to Cambridge, England, specifically to work on the structure of DNA, found these clues interesting. Trained in genetics, he was collaborating with Francis Crick, a British biophysicist who shared his fascination with DNA. Very little was then known about DNA's vital statistics—its diameter, length, density, chemical bonds, or the angle at which it twisted. About the only information one could rely on came from X-ray diffraction patterns. These showed the distance between the nucleotides in DNA and pointed to a large repeated structure of some kind. But the X-ray patterns did not reveal how many strands of nucleotides the DNA molecule contained, nor whether the backbone of the structure was on the inside or outside of the strands.

Being short of facts, Watson and Crick resorted to building models of what the DNA molecule might look like and then testing these models. Did every precisely scaled piece of the model fit what was known from the X-ray data? Did it obey standard rules about chemical bonds? This was an approach they had learned from Linus Pauling of the California

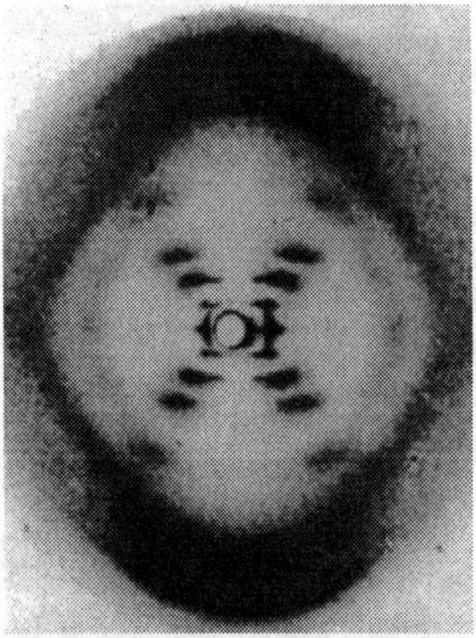

The X-ray diffraction photograph of DNA taken by Rosalind Franklin that was published with the paper by Watson and Crick in *Nature*, April 25, 1953.

Institute of Technology, who used it very successfully in discovering the shapes of some protein substructures (see page 22).

Their first models failed the tests. But eventually—after learning more about DNA from some new X-ray pictures taken by Rosalind Franklin at King's College—they devised a double helix whose basic structure repeated itself every 34 hundred-millionths of a centimeter (34 angstroms), exactly 10 times the distance between one nucleotide and the next. They put two backbones on the outside and had the nucleotides meet in the center.

While shifting around cardboard cutouts of nucleotide bases (key components of nucleotides), Watson suddenly realized that one of these bases, adenine (A), could form two hydrogen bonds with another one, thymine (T). Moreover, the bond lengths were correct for the model. Another base, guanine (G), could make similar hydrogen bonds with the fourth base, cytosine (C). When Watson compared the two cardboard pairs (A with T and G with C), they turned out to form nearly identical shapes that could fit snugly inside the backbones. This conformed with—and explained—a finding made 3 years earlier by Erwin Chargaff of Columbia University that the number of A bases in DNA was equal to that of T bases, while the number of G bases was equal to that of C bases.

James Watson and Francis Crick with the first full model of DNA.

From the way the bases attached to the backbones, which consisted of chains of sugars and phosphates, Crick then saw that the two chains must run in opposite directions. The bases could appear in any order on one strand of DNA, but this order determined the sequence of the complementary bases on the other strand. He also saw a built-in means of replication.

"We imagine that prior to [cell] duplication the hydrogen bonds [connecting the base pairs] are broken, and the two chains unwind and separate," Watson and Crick wrote in a paper in *Nature* in 1953. "Each chain then acts as a template for the formation on to itself of a new companion chain, so that eventually we shall have *two* pairs of chains, where we only had one before. Moreover, the sequence of the pairs of bases will have been duplicated exactly."

Watson and Crick also suggested that "the precise sequence of the bases is the code which carries the genetical information." This meant that the four bases were a kind of alphabet—a small set of "letters" with which an infinite number of instructions could be written.

The instructions contained in DNA sequences tell the cell how to manufacture the thousands of enzymes and other proteins on which life depends. But these instructions are not transmitted directly. The DNA remains safely in the nucleus, somewhat like the printing block in a printing press. Meanwhile a copy of one strand of DNA is made in the nucleus, leaves it, and directs the production of proteins in other parts of the cell. This "working copy" of the DNA is a nucleic acid called RNA.

The language in which the DNA's and RNA's instructions are transmitted—the genetic code—was deciphered in the 1960's by Marshall W. Nirenberg of the National Institutes of Health (NIH), Severo Ochoa, who was then at New York University, and Har Gobind Khorana, who was then at the University of Wisconsin, a discovery for which Nirenberg and Khorana later won the Nobel Prize. The code is based on triplets of nucleotides, or "codons," which are read in sequence. Each codon specifies either one of the 20 amino acids or a signal to start or stop constructing an amino acid chain. Each gene consists of a series of codons that contain the instructions for building a specific protein, which influences a specific trait, or that make the RNA used to carry out the DNA's instructions.

For years scientists had thought that nucleotides occurred in regular, repeating sets. Watson and Crick's model made it clear that nucleotides occur in infinitely varied sequences. There are 3 billion pairs of nucleotides in the DNA of a typical human cell, and each person's genetic material has a unique nucleotide

The genetic code. Each triplet of nucleotides codes for an amino acid, except for three–the "stops"–which signify the end of a protein chain. One amino acid, methionine, can also act as a signal to start protein production.

UUU	Phe	UCU	Ser	UAU	Tyr	UGU	Cys
UUC	Phe	UCC	Ser	UAC	Tyr	UGC	Cys
UUA	Leu	UCA	Ser	UAA	stop	UGA	stop
UUG	Leu	UCG	Ser	UAG	stop	UGG	Trp
CUU	Leu	CCU	Pro	CAU	His	CGU	Arg
CUC	Leu	CCC	Pro	CAC	His	CGC	Arg
CUA	Leu	CCA	Pro	CAA	Gln	CGA	Arg
CUG	Leu	CCG	Pro	CAG	Gln	CGG	Arg
AUU	Ile	ACU	Thr	AAU	Asn	AGU	Ser
AUC	Ile	ACC	Thr	AAC	Asn	AGC	Ser
AUA	Ile	ACA	Thr	AAA	Lys	AGA	Arg
AUG	Met (start)	ACG	Thr	AAG	Lys	AGG	Arg
GUU	Val	GCU	Ala	GAU	Asp	GGU	Gly
GUC	Val	GCC	Ala	GAC	Asp	GGC	Gly
GUA	Val	GCA	Ala	GAA	Glu	GGA	Gly
GUG	Val	GCG	Ala	GAG	Glu	GGG	Gly

U – uracil (thymine)
C – cytosine
A – adenine
G – guanine
Ala – alanine
Arg – arginine
Asn – asparagine
Asp – aspartic acid
Cys – cysteine
Gln – glutamine
Glu – glutamic acid
Gly – glycine
His – histidine
Ile – isoleucine
Leu – leucine
Lys – lysine
Met – methionine (start)
Phe – phenylalanine
Pro – proline
Ser – serine
Thr – threonine
Trp – tryptophan
Tyr – tyrosine
Val – valine

sequence, duplicated only in identical twins.

The sequence of nucleotides in a piece of DNA can now be determined through increasingly efficient techniques, some of which are being automated. Scientists know that these sequences contain the instructions for making proteins. They can "read" these instructions. They also know how to determine the sequence of amino acids in a given protein. But to date nobody understands precisely how the sequence of amino acids leads to the remarkably complex and irregular three-dimensional structures of proteins.

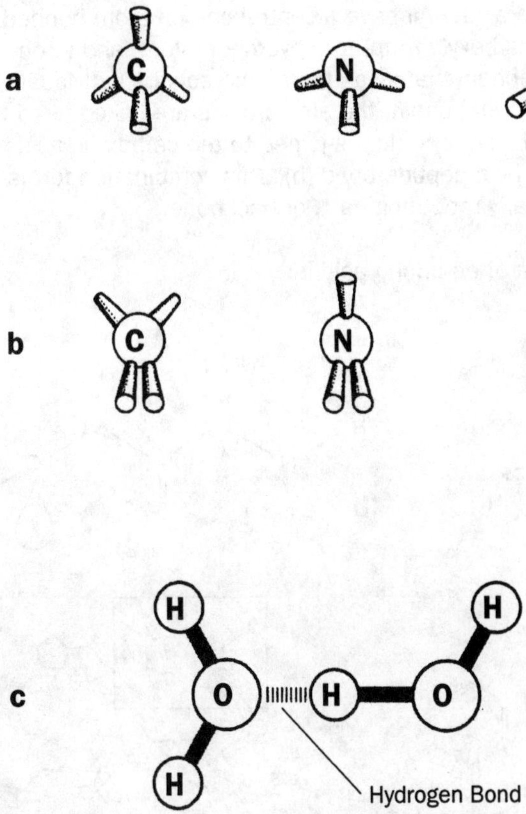

The atoms that make up all matter establish bonds with each other to form molecules. Six elements—carbon, hydrogen, nitrogen, oxygen, phosphorus, and sulphur—make up the amino acids that form proteins and the nucleotide bases that form DNA and RNA. In chemical diagrams, these elements are represented by the first letters of their names.

The strongest and most stable bonds between atoms are covalent bonds. When a covalent bond is single (a), the molecular structure is flexible. When it is double (b), it is more rigid, restricting the molecular structure to a narrower range of possibilities. As a rule, each type of atom forms a characteristic number of covalent bonds with other atoms.

There are also weak bonds that help stabilize molecular structure. Hydrogen bonds (c) are the commonest of these. Many hydrogen bonds form simultaneously as the protein folds, thus "locking in" the shape.

There are 20 amino acids. All have a central carbon atom bonded to an amino group, to a carboxyl group, to a hydrogen atom, and to one variable group—a side chain that gives the amino acid its individuality (a). In a polypeptide (protein) chain, the amino acids are linked head to tail: the amino group (H_2N) of one unit is joined to the carboxylic acid group (COOH) of the next by a peptide bond (b). This combination forms the polypeptide's regularly repeating part, or backbone.

The general formula of an amino acid is:

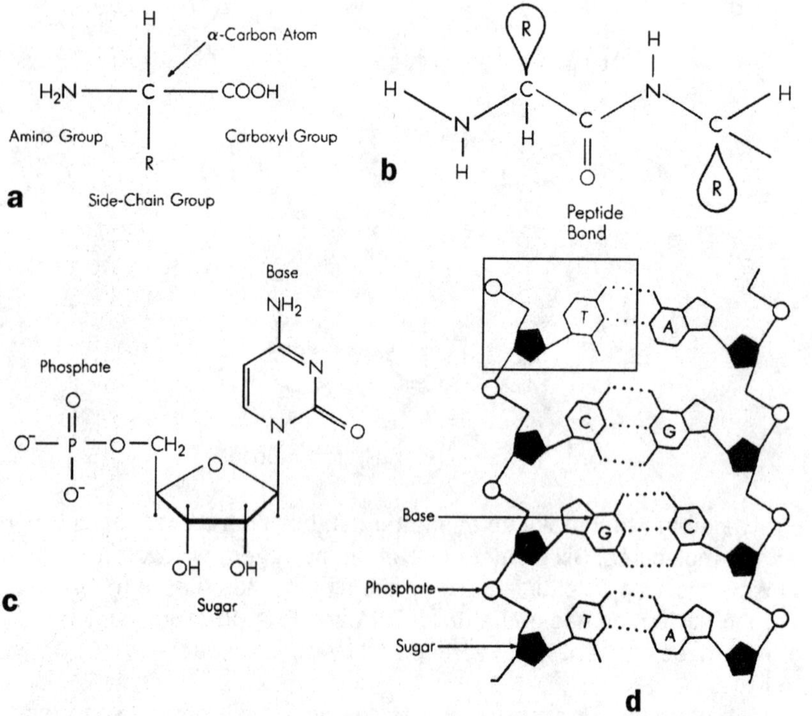

There are four nucleotide bases in DNA: adenine (A), thymine (T), guanine (G), and cytosine (C) (in RNA, the base uracil substitutes for thymine). Each DNA base is attached to a sugar and a phosphate, which form the outer backbone of the molecule (c). An adenine base will fit only with thymine, and a guanine base will fit only with cytosine (d). The linked nucleotides then twist into a very long helix.

DNA transcription. The instructions contained in DNA are transmitted to other parts of the cell by an intermediary — a copy manufactured out of a different nucleic acid, RNA—while the original DNA remains safely in the nucleus. The RNA copy is made in the nucleus by transcribing (producing a new complement to) just one strand of DNA, which is sufficient to encode the instructions.

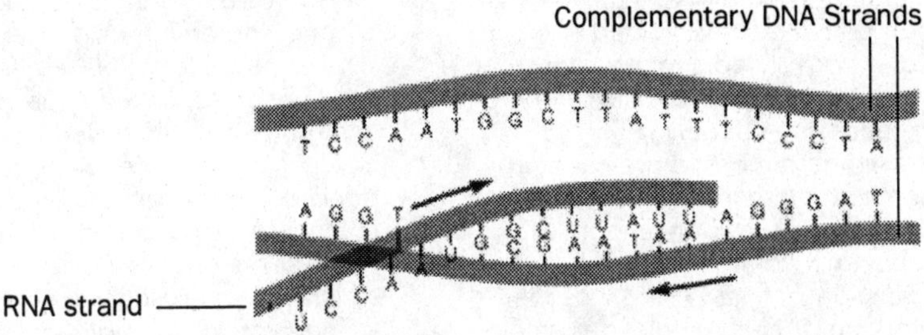

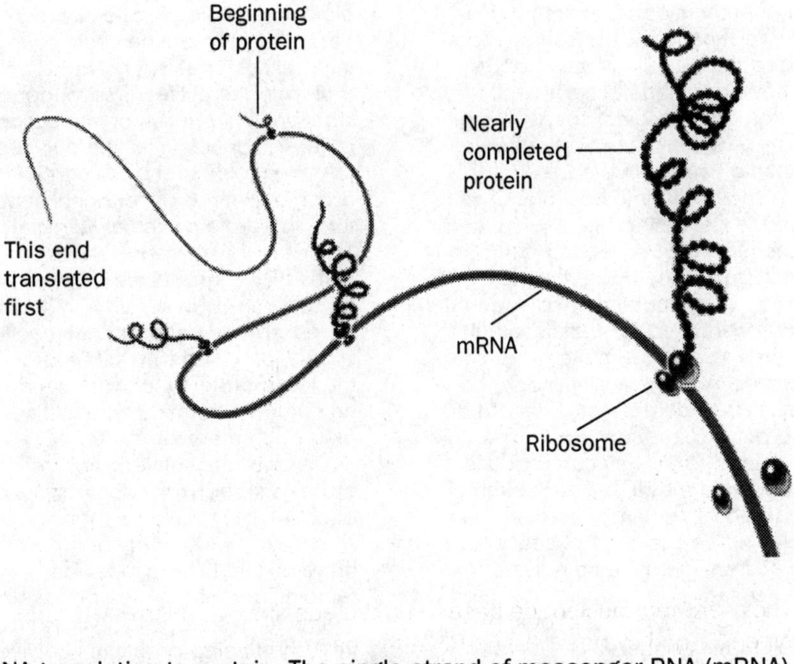

RNA translation to protein. The single strand of messenger RNA (mRNA) carries the DNA's instructions to the ribosomes, tiny protein factories outside the nucleus. The ribosomes translate these instructions into growing chains of amino acids that become specific proteins.

WHY DO PROTEINS FOLD?

A striking characteristic of protein molecules is that they cannot function until they have folded into intricate, convoluted shapes. Only then does their biological activity—and, in a sense, life—begin.

Folding directions are encoded somewhere in the sequence of amino acids. Evidence that such rules exist came in 1961, when Christian Anfinsen, who was then working at NIH, unfolded a small enzyme, ribonuclease, by immersing it in a chemical solution that disrupted the weak bonds that stabilized the protein structure and showed that the protein would refold in its original shape, regaining its power as an enzyme, after the solution was removed.

Once a protein has folded, its shape determines its physical and chemical properties and particularly its active sites, where the protein does its job of binding to—and acting on—molecules that fit with it. When scientists wish to produce a protein with a specific shape, however, they do not know how to do so or what sequence of amino acids will lead to it. They can produce a sequence at will, but very often, instead of creating a well-shaped and active protein, this sequence leads only to a clumpy mess.

When natural proteins are made by living cells, a strand of RNA, called messenger RNA, leaves the nucleus and delivers its instructions to tiny cellular bodies called ribosomes, the "factories" where protein molecules are manufactured. One by one, the amino acids specified by the messenger RNA emerge from the ribosomes, linked head to tail. They form a chain called a polypeptide, which has a regularly repeating backbone and distinctive side chains. (Amino acids differ from one another only by their side chains, which give them individuality.)

The protein's primary structure is this linear chain of amino acids. But almost immediately after it is created, the polypeptide chain springs into helices, sheets, or other shapes that form the protein's secondary structure. As a result of interactions of these various units, the polypeptide then folds and coils further into a complex three-dimensional structure, the protein's tertiary structure. This often brings together amino acids that might otherwise be at opposite ends of the chain. In some cases, several separate polypeptide chains may bind together into a single large protein molecule, such as hemoglobin, forming what is known as a quaternary structure.

These complex interactions give proteins stability; random sequences of amino acids seldom form stable structures. But just why and how proteins fold the way they do remains one of the major mysteries in structural biology. Scientists call this "the folding problem." Solving this problem could produce rapid advances in understanding protein

Linus Pauling with a model of the structure of the alpha helix.

structures, and in synthesizing proteins for commercial and therapeutic uses.

One way to tackle the folding problem is to study the attraction between some of the atoms in a protein. In general, chemical reactions tend to minimize the "free energy"—the stored, potential energy in a molecule—and thereby lead toward a state of equilibrium. Polypeptide chains fold according to the same general plan. So if scientists knew all the forces that act on each atom in a protein, as well as on the surrounding molecules, they would be able to predict the protein's three-dimensional structure on the basis of its amino acid sequence alone. In fact, Harold Scheraga of Cornell University has tried to do this with the aid of computers. But so many low-energy states are possible for the atoms in the molecules specified by a single amino acid sequence that no computer can determine which choice is correct.

Linus Pauling, who had pioneered the study of the elemental forces between atoms in molecules, soon moved on to analyzing the specific structure of an amino acid. First he worked out some of the rules behind this structure—for example, that atoms on either side of a peptide bond must lie in a plane. Then he built models of amino acids based on what he could deduce from X-ray diffraction studies, making sure that each piece of these giant jigsaw puzzles obeyed his new rules.

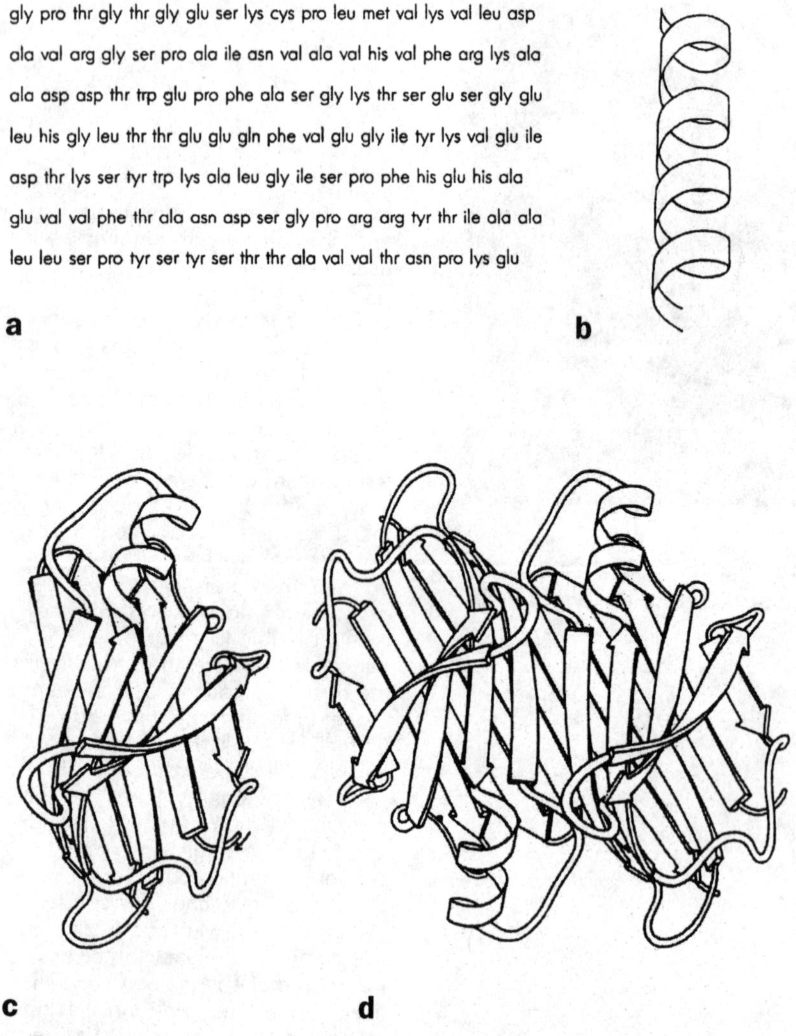

Protein structure. The amino acid sequence (a) is called the primary structure. It coils and folds (b—the secondary structure) and forms a complex 3-D structure, called the tertiary structure (c), which brings together amino acids that might otherwise be at opposite ends of the chain. A quaternary structure (d) is formed when separate chains bind together, forming domains of a single large protein.

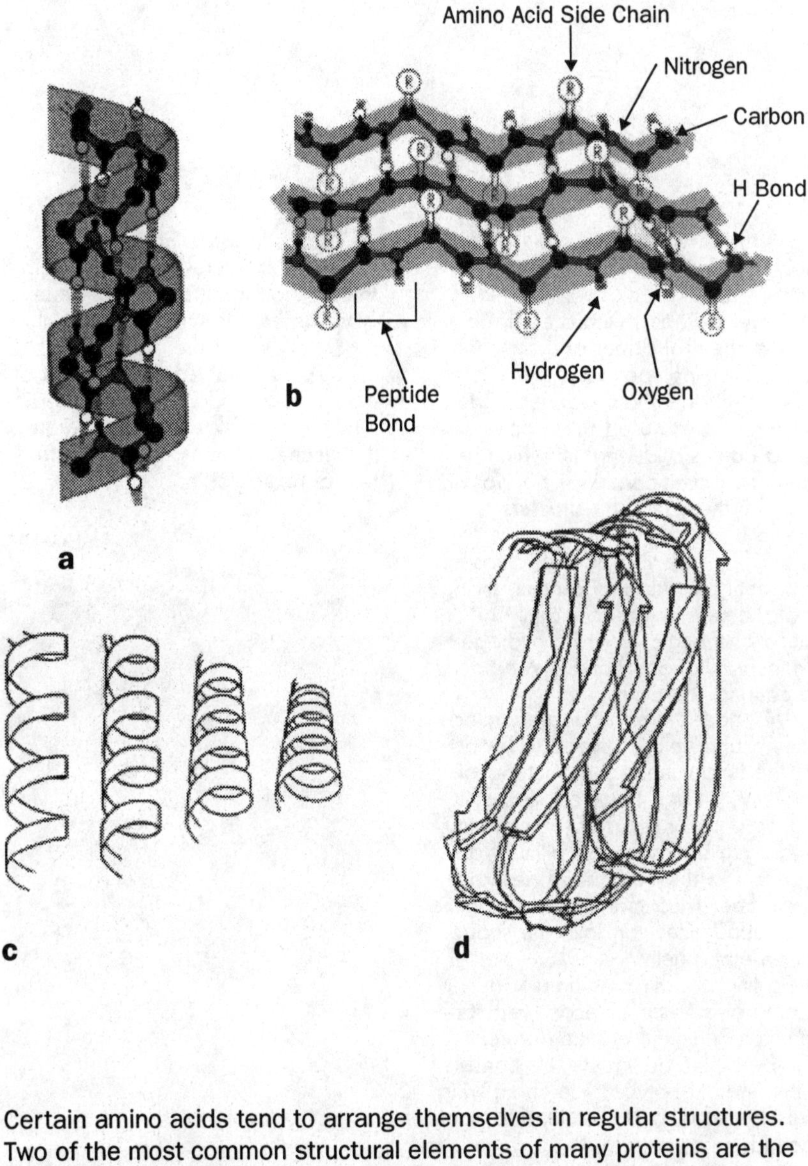

Certain amino acids tend to arrange themselves in regular structures. Two of the most common structural elements of many proteins are the alpha helix (a) and the beta-pleated sheet (b). In simplified drawings, the alpha helix (c) is shown as a coil drawn from varying angles to represent its three-dimensional position in the molecule. Beta-pleated sheet strands (d) are represented by arrows.

What, he wondered, did polypeptides—the chains of amino acids—actually look like? In 1948, while lying in bed with a cold, he took a sheet of paper and sketched out what a polypeptide's structure might be if the chain were spread out straight. He used the proper bond angles and lengths, which he knew from memory. Since he was sure, by then, that the structure would be some sort of helix, he folded the paper in parallel creases and bent the strip into a helix, trying it at different angles. Suddenly he found one angle at which hydrogen bonds would form at the right places to connect the turns.

This model, which he later called the "alpha helix," proved to be one of the fundamental substructures of proteins. Its backbone consists of a tightly coiled helix, but the numerous hydrogen bonds that stabilize it do not involve the side chains, leaving them free. Thus, many different amino acid sequences can take the shape of an alpha helix.

Pauling and his associate Robert Corey also found a second regular structural element of many proteins: the beta-pleated sheet. This consists of strands of polypeptide chain lying side by side, connected by hydrogen bonds.

Despite the growing information about substructures, deciphering the three-dimensional shapes of proteins was—and still is—a lengthy, painful process. One of the earliest and most spectacular efforts was that of Max Perutz of Cambridge University, who struggled for 22 years before he succeeded in mapping the structure of hemoglobin.

DISCOVERING THE "HIDEOUS" SHAPE OF HEMOGLOBIN

Hemoglobin is the protein in red blood cells that carries oxygen from the lungs to the tissues, and also helps transport carbon dioxide on its return trip. Obviously, it plays an extremely important role in the body. Abnormalities in its structure can lead to such life-threatening diseases as sickle cell anemia or thalassemia.

Besides being important, hemoglobin is also one of the most plentiful proteins in vertebrates. In the human bloodstream some 5 billion red blood cells, each containing 280 million molecules of hemoglobin, can be found in each milliliter of blood.

To decipher the structure of this enormous molecule, Perutz had to determine the positions of 10,000 atoms, each one located in space by three coordinates. The 574 amino acids formed by these atoms were strung together in four separate, but connected, polypeptide chains. Perutz grew his own crystals of hemoglobin, took his own X-ray pictures, and did millions of computations without the help of a computer, looking for some regular, underlying features that would simplify his problem. He thought the molecule might have four straight layers, each layer being a polypeptide chain. But he never found anything that clear-cut.

His first break came in 1953 when he and his colleagues invented a method that has been called the "Rosetta Stone" for interpreting X-ray diffraction patterns of large proteins: the use of certain "heavy" metal atoms, which are added to the hemoglobin crystals to serve as atomic markers. These metal atoms are called heavy because they have an extremely large number of protons, neutrons, and electrons, and the advantage of using them is that they scatter X-ray beams very strongly.

Perutz realized that if atoms of mercury or other heavy metals were added to the protein before it was crystallized, they would attach themselves to specific sites in the hemoglobin molecule and produce measurable differences in the intensities of the spots in the X-ray diffraction pattern. By comparing the X-ray patterns of hemoglobin with and without these heavy atoms, he could more precisely determine the intensity and angle of the X-ray reflections. This would allow him to analyze the molecule in three dimensions for the first time.

Six more years of effort were required to make the method work. Finally, all the pieces came together and Perutz stood before the completed structure of the hemoglobin molecule—a bizarre mass of twisted chains with four separate pockets, each one containing the dark red, iron-bearing component called heme that is essential for the binding of oxygen and gives blood its color.

Shocked by the result of his labors, Perutz puzzled about its meaning. "Could the search for ultimate truth really have revealed so hideous and

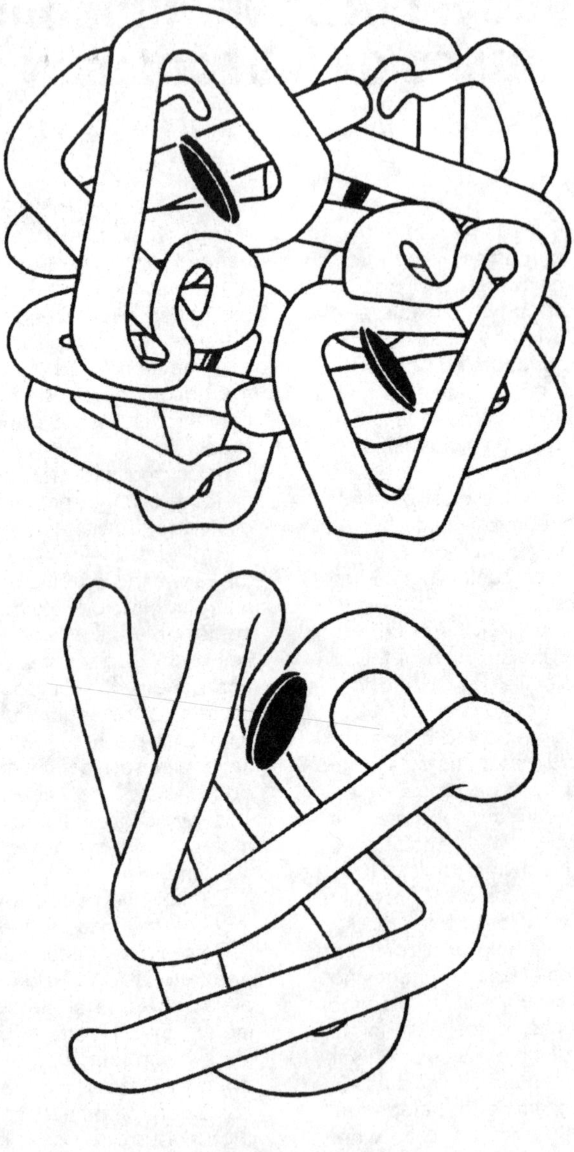

Hemoglobin (a) is one of the most plentiful proteins in vertebrates. Each of hemoglobin's four units looks like one molecule of myoglobin. Myoglobin (b) has one pocket for a heme group, while hemoglobin has four.

visceral-looking an object?" he wondered.

John Kendrew, who had joined Perutz in Cambridge in the 1940's, had a similar reaction after deciphering the structure of a smaller protein, myoglobin, 2 years earlier. Kendrew had chosen to study myoglobin in the hope that it would have a fairly simple structure, but to his surprise the molecule turned out to have a complex, asymmetrical shape.

Myoglobin's job is to take oxygen from the red cells in the bloodstream and store it in muscle tissue until needed. This storage occurs by the binding of an oxygen molecule to the iron atom at the center of myoglobin's heme group. Since diving birds and mammals that spend much time underwater have abundant supplies of myoglobin in their muscles, Kendrew worked with myoglobin from sperm whales. After Perutz devised the heavy-metal method of X-ray crystallography, Kendrew succeeded in solving myoglobin's structure. This was the first three-dimensional model that showed the full complexity of a protein molecule.

At first there seemed to be neither rhyme nor reason to myoglobin's shape, but when Perutz determined the structure of hemoglobin the obvious similarities between the two models began to offer some clues to the significance of their design. Myoglobin has one pocket for a heme group; hemoglobin, four

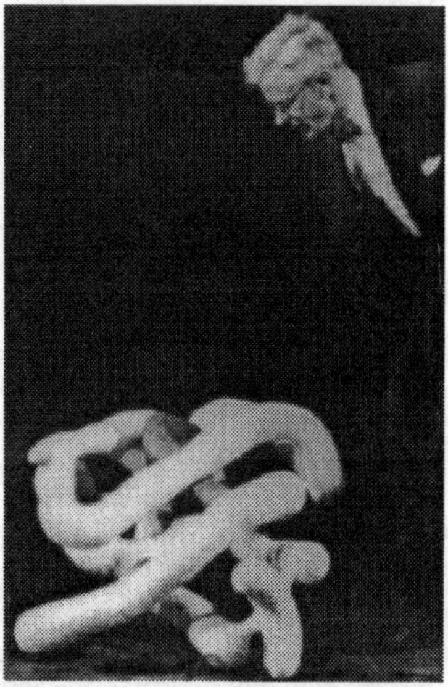

John Kendrew with the model of myoglobin at low resolution. The heme group is the gray disk in the photo.

times its size, has four. Each of hemoglobin's four subunits looks like one molecule of myoglobin. And the interior of all these units consists almost entirely of hydrophobic (water-avoiding) amino acids, while their exterior consists of hydrophilic (water-loving) amino acids. This allows the iron atoms in the buried heme group to bind with oxygen without becoming oxidized (rusted).

In 1962, Perutz and Kendrew won the Nobel Prize in chemistry for developing the heavy-metal method of X-ray diffraction and for solving the structures of hemoglobin and myoglobin.

Since this achievement, scientists have deciphered the three-dimensional atomic structure of well over 500 additional proteins. In fact, the structures of most of the proteins that scientists have succeeded in growing into crystals for study by X-ray diffraction are now known. But only a fraction of proteins, which normally exist in solution, crystallize like sugar or salt when the surrounding water is removed. The others either form an amorphous gel, with their molecules pointing in different directions, or turn into a powder. Little is known about the shapes of these tens of thousands of proteins that do not crystallize, although a new technique called 2-D NMR may soon come to the rescue (see page 39).

Although the amount of data on protein sequences is increasing at an astounding rate because of advances in molecular biology, the folding problem remains a major challenge to scientists. Being able to predict the detailed three-dimensional structure of a protein from a given sequence of amino acids is "the most fundamental problem at the chemistry-biology interface," according to a report by the National Academy of Sciences, "and its solution has the highest long-range priority."

RECOGNIZING PROTEIN STRUCTURES

At first glance, protein structures seem disorderly and incomprehensible. Yet it has become increasingly clear that many proteins do share some basic elements and motifs.

One way to find these common features is to survey all known protein structures "as though we have been presented with the tropical jungles of a totally new planet," says Jane S. Richardson of Duke University. Using this approach in 1976, Michael Levitt, who was then at the Weizmann Institute of Science in Rehovot, Israel, and Cyrus Chothia of the Medical Research Council in Cambridge, England, showed that the proteins whose structures were then known could be assigned to one of five structural classes, depending on the pattern in which their amino acid chains are folded. Richardson further analyzed these patterns, made elegant schematic drawings that reveal the underlying similarities of many proteins, and created a taxonomy of protein shapes.

Two of these structural components—the alpha helix and the beta-pleated sheet—were first identified by Linus Pauling in 1951. Richardson pointed out that since alpha helices are particularly stable, a single helix can have more influence on the stability and organization of a protein than could any other individual structural element. Perhaps because the helical structure is so vital to stability, very few proteins are known to have no helix whatsoever. Furthermore, many proteins are made up largely of helical structures.

Another common feature of protein structures is the "tight turn." These turns generally occur at the surface of a protein; their role is to interrupt or join together structures that are more internal, and they exert a strong influence on the process of protein folding.

According to Richardson, tight turns and "beta bulges"—irregularities in certain beta sheets—often indicate the presence of a protein's active site.

Apparently, the same structural elements occur over and over again in different proteins. In 1974, Michael Rossmann of Purdue University identified a recurring pattern in many proteins that bind to nucleotides: two beta strands connected by a cross-over alpha helix. He called it the "mononucleotide-binding fold." Surprisingly, this fold has been found in proteins that bind to nucleotides even when the amino acid sequences of these proteins appear totally unrelated.

As more and more structural similarities are revealed, researchers are moving up to higher levels of organization. They have identified about 100 different "domains" or "structural motifs," each one made up of several basic elements and forming a compact cluster that looks somewhat like a complete small protein. Richardson has classified these

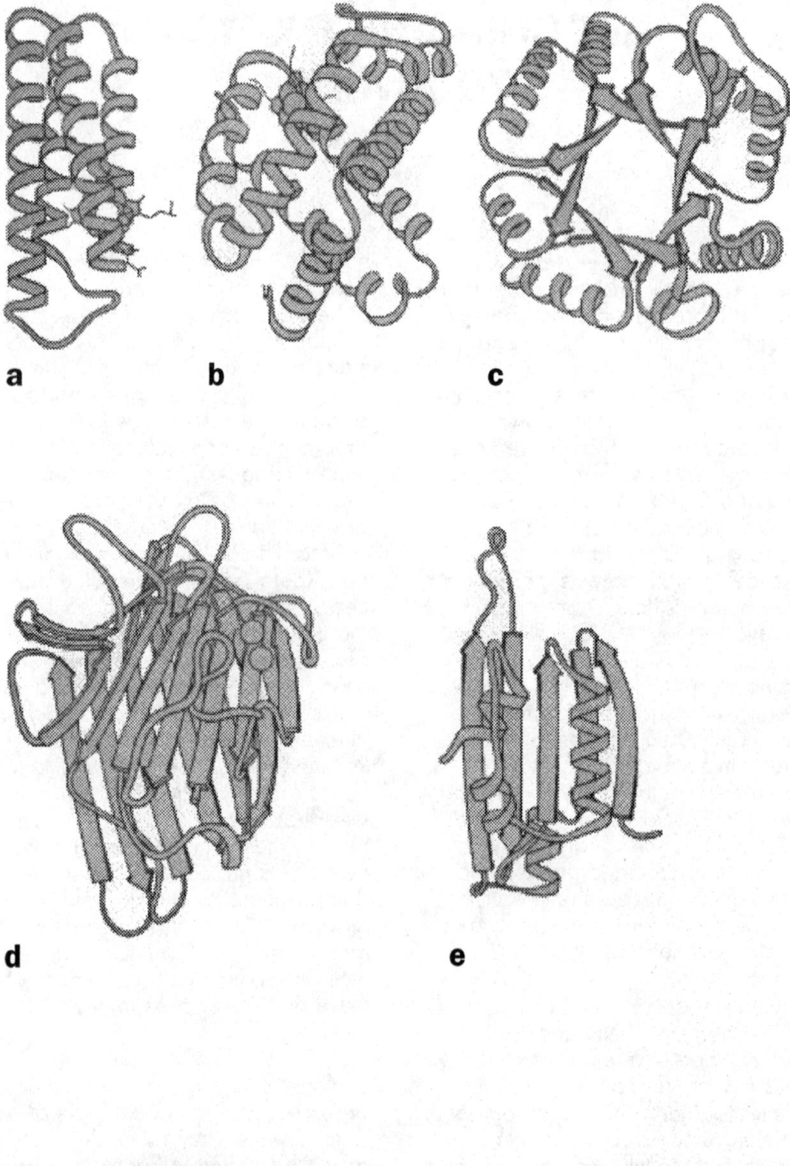

Some of the classifications given to protein structures by Jane Richardson include: up-and-down alpha helix bundles (a), Greek key helix bundles (b), beta barrels (c), jellyroll Greek key beta barrels (d), and open-face beta sandwiches (e).

domains under such headings as "beta barrels," "jellyrolls," "Greek keys," and "open-face beta sandwiches."

Not only do these domains make pretty pictures, but they seem to carry out specific functions. Most significantly, as Rossmann has pointed out, the active center of a molecule is often at the interface between domains, where the simple functions of each domain are brought together to form a more sophisticated molecule.

Rossmann has even suggested that primitive proteins may have been created by the fusion of useful domains of this sort. This hypothesis fits in with the discovery, in 1977, that mammalian DNA sequences that code for genes are interspersed with "intervening sequences," or "introns," of unknown function. In some cases, introns have been found between the sequences that code for separate domains.

Walter Gilbert of Harvard University believes that the coding sequences of DNA (now called "exons") represent the "movable domains" hypothesized by Rossmann. He argues that "exon shuffling" allows the reassembly of domains in new combinations, which encourages the evolution of potentially useful proteins.

Can a protein's domains be predicted from its amino acid sequence? Some scientists think so, and they are looking for clues in the Brookhaven National Laboratory's Protein Data Bank, a repository of known 3-D structures and sequences, as well as in the Protein Identification Resource, a repository of protein sequence information at Georgetown University in Washington, D.C. The information in these banks is growing much more rapidly than is the ability to analyze it. Determining the amino acid sequence of a given protein used to be extremely complicated, but the procedure was automated in 1970. Since then, even this method has been largely supplanted by a more rapid, though indirect, approach based on recombinant DNA technology. In this approach, scientists first determine the nucleotide sequence of a stretch of DNA that directs the making of a protein, and then translate this sequence into the amino acids for which it codes.

Several groups have attempted to develop methods for predicting that certain sequences of amino acids will form alpha helices, others will form beta strands, and yet others will form turns. More complex search schemes based on artificial intelligence have also been brought into play. A number of these methods show promise, although none is perfect. In addition, scientists are attempting to predict which of the basic elements of proteins will combine into larger units, or even into specific domains.

"We search the database and flag every place there might be a beta sheet or an alpha helix or a

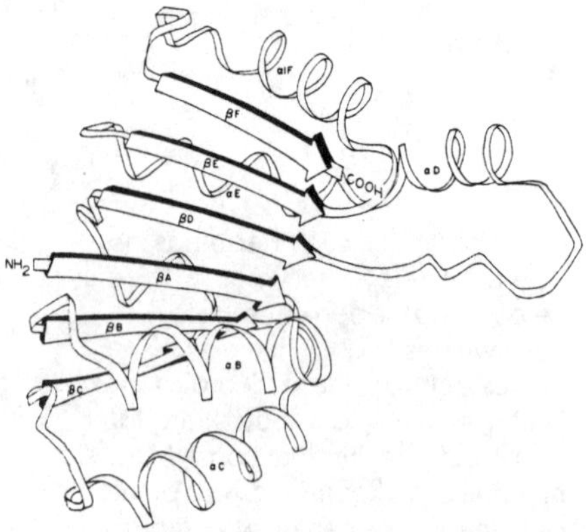

Sketch of a mononucleotide-binding fold showing alpha helices and beta-pleated sheets.

hydrophobic [water-avoiding, therefore internal] area,'' explains Temple F. Smith of the Dana-Farber Cancer Institute in Boston. ''Then, using a scheme developed at the Artificial Intelligence Laboratory at MIT [the Massachusetts Institute of Technology], we can identify probable functional domains, such as the mononucleotide-binding fold.''

Evidence that this approach works came when the Dana-Farber/MIT group examined the amino acid composition of a protein from a virus that has been linked to bladder cancer. They were able to predict, correctly, that part of the protein would be in the shape of the mononucleotide-binding fold. This led to experimental predictions that were tested by modifying specific amino acids in the protein.

The goal of such studies is to learn to recognize how certain amino acid sequences determine specific protein shapes and functions—to decipher what has been called ''the second genetic code.'' Smith believes that ''we are now beginning to crack this code.''

PROTEIN DYNAMICS

A decade ago, most scientists believed that protein molecules had to fit together like a lock and key in order to perform their functions. It was therefore assumed that proteins had rigid, fixed structures. Although there were those who argued that proteins were more flexible, the wire models of protein structures derived from X-ray crystallography provided a strong visual reinforcement for the concept of protein rigidity.

One protein, at least, had been caught in the act of changing shape as far back as 1961: hemoglobin. A scientist in Perutz' laboratory had discovered that every time the hemoglobin molecule took up oxygen, the iron atoms in two of its chains moved into the plane of the heme. This motion triggered a whole series of structural changes in the molecule. The changes sent signals about the binding of oxygen to all parts of the molecule, thereby increasing the molecule's affinity for additional oxygen. When the hemoglobin molecule released oxygen, it returned to its original shape.

Perutz had been struck by this regular activity. He called hemoglobin "a kind of molecular lung. . .an organ in miniature." Nevertheless, since proteins were believed to be rigid, scientists assumed that hemoglobin simply had two rigid forms and that the molecule underwent an abrupt transition from one rigid form to the other.

Martin Karplus, a Harvard physical chemist who was at the forefront of the revolution to a more dynamic view of proteins, explains that such "misconceptions" arose because "when you do X-ray diffraction, out comes a picture in which every atom appears to be fixed in place." But in reality, what X-ray crystallographers examine is "an average structure, one that does not exist," he says. He points out that a crystal that is prepared for X-ray analysis may consist of trillions of protein molecules, "yet it is highly improbable that at any instant of time even one individual protein molecule has the average structure."

As scientists now recognize, the atoms in a protein molecule are in a state of constant motion. They fluctuate around an average structure, even within molecules that are neatly lined up in a crystal.

These atoms move around inside the molecules because of thermal energy. At room temperature, the atoms in proteins undergo significant fluctuation.

Various aspects of protein structure limit the atoms' freedom to move. Pairs of atoms may be held together by strong bonds, for instance; nevertheless, they are free to twist—this is what makes polypeptide chains flexible. As the chain twists and turns, its various side chains move with it, much as happens in mobiles by the sculptor Alexander Calder. The side chains also have bonds around which rotation can occur.

"Owing to the tight packing of the protein's interior, a typical atom cannot move far unless its neighbors move also," Karplus points out. "Nevertheless, the combination of many small local motions can produce large-scale, more global displacements of one part of a protein with respect to another." Some scientists refer to this larger movement as the protein's "breathing."

A new discipline—molecular dynamics—examines how the movement of atoms inside a protein molecule influences the protein's activity. This involves calculating the acceleration of each atom at a particular instant, according to Newton's laws of motion (see glossary), while taking into account the spring-like force of the chemical bonds between atoms (potential energy) and the forces between nonbonded atoms, including those that prevent more than one atom from occupying the same point in space at any one time.

"In a protein consisting of many atoms, the total force acting on any one atom at any given time depends on the positions of all the others," Karplus explains. "Not surprisingly, . . .determining the positions and velocities of all the atoms in a protein requires a high-speed computer. Such calculations constitute what is called a molecular-dynamics simulation."

Computers, in fact, are at the heart of this new approach. Starting with a set of atomic positions derived from X-ray crystallographic data, scientists can now calculate the distance that each atom in a molecule has traveled in an infinitesimally short interval, or "time slice." Then they can take successive "snapshots" of the protein structure, tracing the individual paths of all the atoms in the molecule over a period of up to a million such time slices (which typically would add up to one nanosecond, or a billionth of a second).

Many major motions of a protein can take place within this time frame. And these motions are essential for protein function.

For example, if the myoglobin molecule were static and looked like Kendrew's model at all times, there would be no way for oxygen to get inside it—and no way for our muscles to store the oxygen they need. It would take billions of years for oxygen to reach down to the heme group in the model's tightly packed center, and even longer for it to bind with the heme and get out again.

What actually happens is quite different. As computer simulations have shown, oxygen can reach the heme group through a path opened up when several side chains of the molecule are rotated out of the way, with only a small expenditure of energy.

Until the 1960's, scientists who wanted to visualize large molecules in three dimensions had to build mechanical models out of rods, balls,

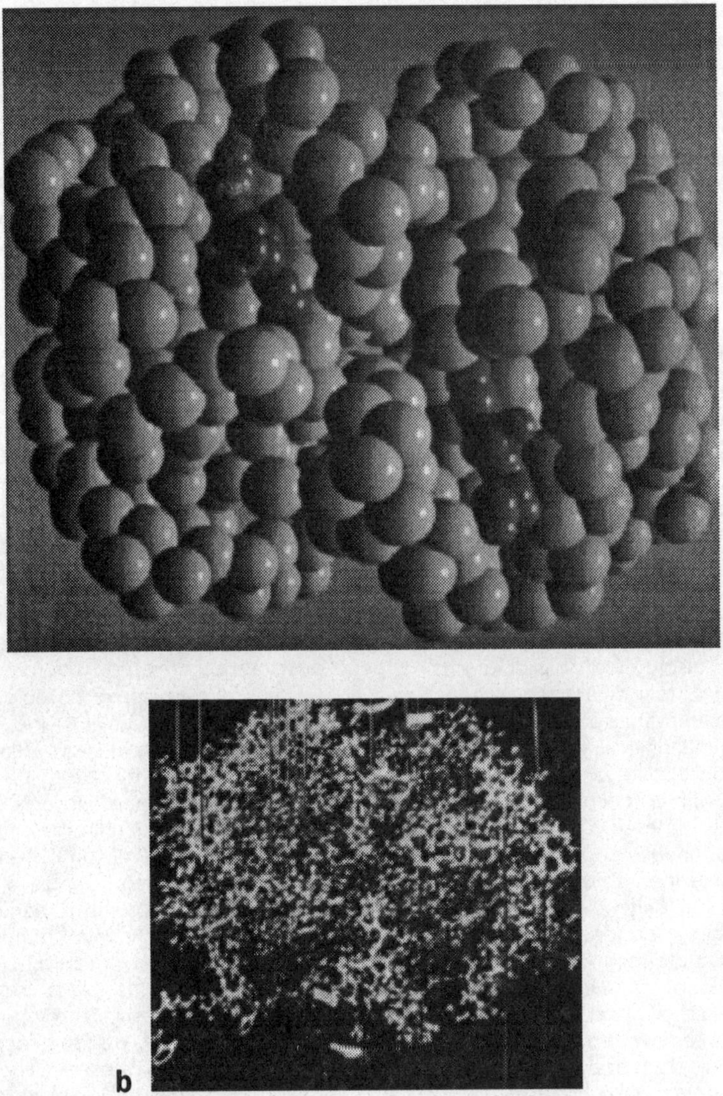

A space-filling computer graphic model of hemoglobin (a). This method of modeling allows researchers to turn the image around and enlarge, reduce, or highlight it; to view it from any angle in its entirety; or to view it in great detail. Computer graphic models have largely replaced the older wire models, such as this one of hemoglobin (b).

and wire scaffolding. It was a clumsy, slow, and painstaking business, and the models often fell apart. Today, all this has been replaced by computer graphics, a technology that was combined with molecular modeling in 1964 at MIT, when Cyrus Levinthal and Robert Langridge (who was then at Harvard) developed the first system that showed biological molecules on a computer screen.

Right from the start, computer graphics allowed scientists to move rigid molecules in space and to alter selected parts of them. The graphics systems rapidly became so powerful and so versatile that researchers could display and manipulate as many as four different molecules on the screen simultaneously, and see their interactions. Today, scientists can dissect the components of proteins with an accuracy that was impossible even 5 years ago. In a sense, molecular graphics has become the ultimate microscope.

But it is only now that the technology has expanded sufficiently to incorporate the moment-to-moment calculations of molecular dynamics. For example, Karplus and his coworkers were able to calculate the trajectory of every one of the 454 atoms in the 58 amino acids of a small protein, the bovine pancreatic trypsin inhibitor, for 100 trillionths of a second, and to show snapshots of this protein's dynamic cycle on a computer screen. This allowed the scientists to follow the protein's small, rapid fluctuations, though not its larger motions, which occur less frequently. These larger motions are more interesting because they determine biological activity. However, tracking them would require more computer time or power, as well as a more sophisticated computer program—all of which are only now becoming available.

In one of the Nation's most powerful computer graphics installations, at the NIH Division of Computer Research and Technology (DCRT), researchers have put together a system that will allow them to take full advantage of this approach.

Sitting at a computer console, Richard Feldmann, chief of the DCRT molecular graphics section, demonstrates the system by calling up a protein on the screen. He puts on a pair of stereoscopic glasses and hands another pair to an observer. Suddenly the protein springs to life in three dimensions on the screen. As Feldmann moves a control handle, parts of the protein structure begin to twist sideways. Other parts zoom forward for a closeup. The whole protein rotates 180 degrees, and a cavity appears in its center. The 3-D illusion is so powerful that one can seem to put one's finger right into the cavity on the screen and explore its boundaries, seeing exactly where another molecule might "dock" and fit inside it.

Computer specialist Richard Feldmann has created a powerful computer graphics system that enables him to study three-dimensional models of molecular structure.

The NIH molecular modeling system uses a software package developed by Karplus' group at Harvard and a specialized computer, called an array processor, that can do 100 million operations per second, Feldmann explains. "Our machine has the power of a supercomputer; it is equivalent to 150 of the typical . . . lab computers," he says.

This system may soon allow scientists to visualize some of the larger internal motions of proteins. "We're not yet at the point where we can calculate large changes over time," says Feldmann. "But we can do in 1 week what it would take a typical lab computer 3 years to do, so we can start a lot of experiments that were not practical before."

A totally different way of studying the motions of proteins and nucleic acids is to analyze the molecules while they are actually free to move—in their natural state, in solution. This is now becoming possible, at least for smaller molecules or for components of larger ones, thanks to a technique called nuclear magnetic resonance spectroscopy.

NMR—and especially a form called two-dimensional (2-D) NMR—is rapidly proving itself as a valuable complement to X-ray crystallography. Not only can it be used to study the thousands of molecules that will not form crystals, but it can capture some of a molecule's dynamics.

NMR works by measuring the radio-frequency waves that are absorbed by the nuclei of atoms when they are placed inside a magnetic field. These nuclei behave like individual magnets, aligning themselves either with the magnetic field or against it. When a short radio pulse of a specific frequency reaches them, a number of these nuclei absorb the radio waves and jump to a higher energy state. Analyzing the spectra produced by these absorbed or "resonant" frequencies provides information on the atomic nuclei's positions in a molecule and on the kinds of atoms that surround them.

Since NMR involves the use of a magnetic field, rather than X-rays, and does no harm to patients, it has become an important diagnostic and research tool. Renamed magnetic resonance imaging (MRI) when used for diagnostic purposes in patients (to remove the ambiguity of the word "nuclear," which has nothing to do with radioactivity in this case), the technique has proved especially useful for investigating the chemistry of the brain and other hard-to-reach organs. In studies of human beings, MRI "listens" to the response of the body's abundant hydrogen atoms, which can be translated into 3-D images of tissues in the body. This provides information on how well body cells are functioning.

The recent development of 2-D NMR is enabling researchers to use the technique for the analysis of biological molecules. "Two-dimensional

Scientists using an NMR spectrometer. The researcher on the left injects a sample protein while the researcher on the right adjusts the pressure in the chamber.

NMR has revolutionized the field," says Frederic M. Richards of Yale University. "It's our only hope for getting a detailed, high-resolution structure of proteins that don't crystallize."

So far, the structures of several dozen smaller proteins have been determined using this technology, according to Adriaan Bax, a physicist at NIH's National Institute of Diabetes and Digestive and Kidney Diseases who has been working to improve 2-D NMR techniques. The size of proteins that can be studied in detail by 2-D NMR is limited to about 100 amino acids; for larger proteins, the spectra become too complex.

Another problem is that collecting information with 2-D NMR may take an hour or several days, while molecules may change shape radically in milliseconds. However, it is becoming possible to arrest the motion at certain points in time, or to label some of the intermediate stages so that they can be reconstructed later. Richards believes that "in the future, 2-D NMR will give us the possibility of detecting the multiple structures of proteins—to the extent that you have any stable structure at all. We'll be able to look at these structures in solution and then add substances that bind to these structures, and see what is happening."

For smaller proteins, 2-D NMR has already matched the accuracy of X-ray crystallography. In 1985, Robert

Huber, a crystallographer at the Max Planck Institute for Biochemistry in Martinsried, West Germany, challenged one of the pioneers of 2-D NMR, Kurt Wüthrich of the Institute for Molecular Biology and Biophysics in Zurich, Switzerland, to analyze the structure of a yet-unsolved protein by 2-D NMR while he did it by crystallography—and then to compare the results. Their battleground was tendamistat, a protein that reduces the level of the sugar glucose in the blood after a starchy meal and could be useful in the treatment of obesity or diabetes. When the contending laboratories sent in their results, obtained by such different methods, the two structures were found to be nearly identical.

"Neither saw the other one's report before sending in his results," says Bax. "Yet the main difference in them was that the X-ray crystallograph showed an ordered structure for the terminal region, while the NMR didn't. In solution the terminal region of this protein is just flopping around, moving, so you can't see it. The NMR structure shows you the protein conformation in solution, which is the relevant structure." However, Bax adds, "comparisons between X-ray crystallographic results and NMR solution structures obtained for a number of small globular proteins indicate that the differences between crystal and solution structures tend to be quite small."

Bax has recently developed new approaches that rely on three- and even four-dimensional NMR, which can greatly alleviate 2-D NMR's size restriction. Other scientists are studying larger molecules with NMR by using variant forms of certain atoms, typically carbon or nitrogen. Still other researchers are trying to combine 2-D NMR with techniques such as molecular dynamics.

DECIPHERING THE STRUCTURE OF MEMBRANE PROTEINS

The billions of cells that make up the tissues and organs in our bodies are constantly bombarded by messages from neighboring or distant cells, as well as from the world outside.

In order to function properly, cells must have a way to recognize which of these messages are meant for them—and to respond appropriately. For this they depend on a thin, delicate fluid membrane that surrounds and protects each cell, much like the skin that covers our bodies.

This surface membrane is interspersed with specialized proteins that act like sense organs. They are the cell's "receptors." They do their jobs by selectively binding to hormones and other molecules, and their activity depends on how well their shapes complement the shapes of these molecules.

"Virtually all the activities of proteins can be understood in terms of such selective chemical binding," explains Russell Doolittle. Many questions about the effectiveness of drugs against diabetes, mental disorders, heart disease, hypertension, and other ills could be answered if we knew precisely how various membrane proteins are constructed and how drugs fit them. But until recently these proteins have been very difficult to analyze because they are not water-soluble and could not be induced to form the crystals that are required for X-ray studies.

The reason these proteins are not water-soluble is related to the structure and function of the membranes in which they are embedded. Membranes such as the cell surface membrane serve to create a separate compartment in which biochemical processes can take place. Since cells exist in a watery environment and also contain a large amount of watery fluid, membranes need to be pretty water-tight. Nature's answer is to make membranes out of phospholipids—fatty substances that do not let water in. In order to remain in their proper positions, membrane proteins also must have sections that are insoluble in water.

A few years ago, a group of West German scientists succeeded in making crystals of a membrane protein by treating the membrane with a detergent that made the protein soluble in water. Soon afterward, the same group deciphered the 3-D structure of a membrane protein at high resolution. This provided the first clear view of the individual atoms in a membrane protein.

In late 1990, two groups of scientists—led by Stephen Harrison at Harvard University and Wayne Hendrickson at Columbia University—independently determined the 3-D structure of the part of the receptor protein on the surface of certain immune system cells through which the AIDS virus enters the body. This work should greatly increase understanding of how the virus works and provide information of use in the design of drugs to block infection.

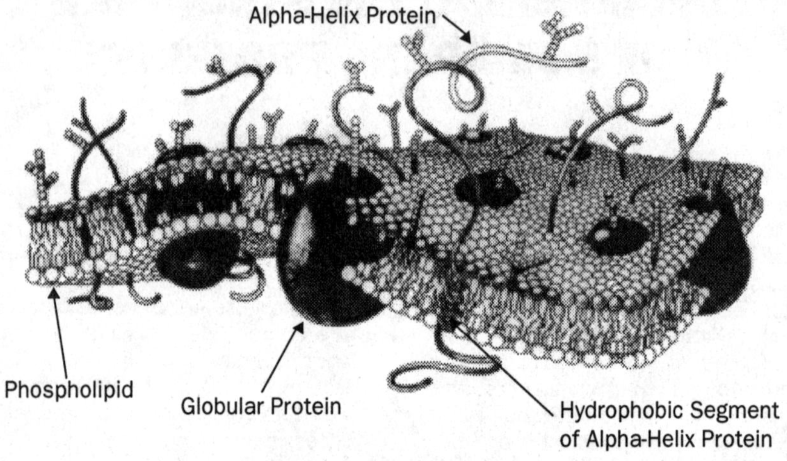

Biological membranes create a separate environment within which biochemical processes can take place. In addition, they control everything that goes into or out of the cell. Membranes consist of two layers of phospholipids with proteins embedded in them. The proteins serve as receptors for specific hormones, neurotransmitters, and antibodies, as well as pores or gates for substances entering or leaving the cell. (The ratio of phospholipid to protein is much greater here than in a natural membrane.)

It is also possible to use NMR to examine the structure of membrane proteins. A leader in this field, Stanley Opella of the University of Pennsylvania, recently was able to describe the structure and dynamics of several small membrane proteins in considerable detail by utilizing a combination of NMR experiments. One technique he employed, solid-state NMR, is especially well-suited for studies of proteins embedded in cell membranes. Opella notes that unlike NMR done on molecules in solution, solid-state NMR has no fundamental limitations to the size of proteins that can be studied, so advances in instrumentation will make it feasible to examine receptors and other large membrane proteins.

HOW PROTEINS TURN GENES ON OR OFF

Each of the 100 trillion cells in a human body contains, in its nucleus, a complete set of all the genes that make a human being—yet only a few of these genes are turned on at any particular time or location in the body. And although each human being starts life as a single cell, which divides again and again, some of these cells become part of an eye, some become part of bones or blood, and others become part of nerves or skin tissue, as commanded by the genes. What controls the activity of each of these genes?

Biologists are just beginning to find some answers to this fundamental question.

Genes have two tasks: to duplicate themselves before cell division, so that each daughter cell has a full complement of genes in its nucleus; and to selectively transcribe their information onto new strands of messenger RNA, which will direct the manufacture of new proteins inside the cell.

Considering the large number and variety of cells whose activity needs to be coordinated at any given time, the sweeping changes that occur in the body as it develops from conception to adulthood, and the many stresses produced by the outer environment, the regulation of the 50,000 to 100,000 genes in our cells seems like a staggering job.

In the past few years, researchers have realized that this job could never be done if the DNA molecule were as static as it appeared to be in the Watson-Crick model of the double helix. Far from being stiff and inflexible, DNA's shape is now recognized as being just as dynamic as that of proteins.

The illusion that the DNA structure was static arose from the fact that scientists were using X-ray diffraction and looking at average structures, just as happened in the study of proteins. The Watson-Crick model of DNA was not even based on X-ray diffraction of crystals, but on a less accurate technique called fiber diffraction. The first crystallographic determination of the double helix at high resolution, showing the exact position of its atoms, was not made until 1973.

Since then, scientists have learned that the DNA molecule has at least three basic forms (see opposite page). In addition, when DNA binds with certain other molecules it can do what one scientist called "very peculiar and unexpected things," such as forming a variety of kinks and bends that change its shape both locally and through a wide stretch of the structure.

Studies of DNA fragments in solution have shown that some of the local bends are permanent. They depend upon—and appear to reflect—specific patterns of nucleotide sequences. More transient changes in DNA structure reflect the molecule's activity as it binds with other molecules to make messenger

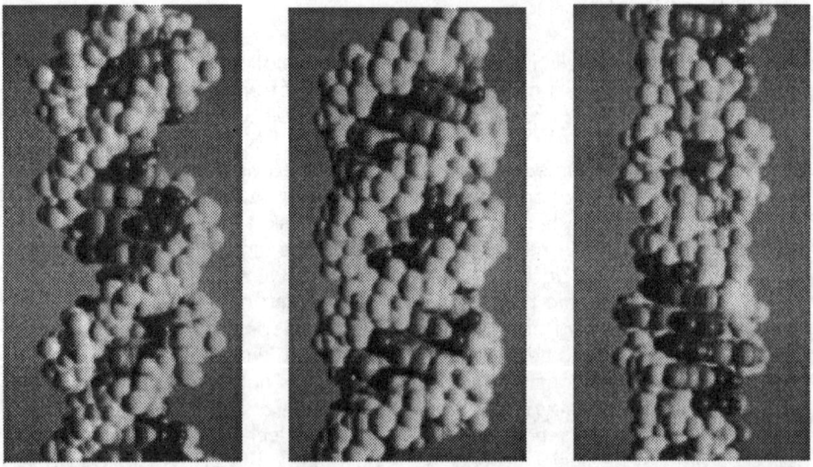

"B" "A" "Z"

Three basic forms of DNA: the right-handed "B" form of the Watson-Crick model; the "A" form, which is a shorter and squatter double helix; and the left-handed "Z" form.

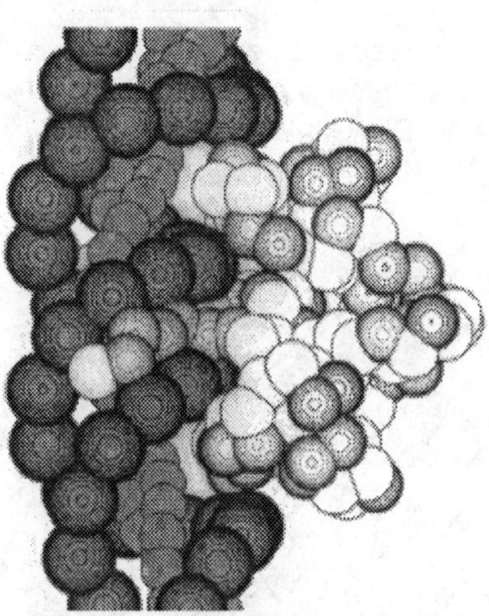

The *cro* repressor (above) bound to DNA.

RNA or to replicate itself.

"DNA doesn't just remain static when proteins sit on it," explains Evangelos Moudrianakis of The Johns Hopkins University, who has been analyzing the interaction of DNA and certain proteins, called histones, around which the DNA is wound. "DNA and proteins talk to each other, and there is a special code for the language that these molecules use together."

Just how individual proteins recognize and bind to specific parts of the giant DNA molecule is the focus of intense research because it holds the key to the control of DNA activity. If scientists can uncover these mechanisms, they may be able to cure certain developmental disorders and other conditions in which DNA is activated at the wrong time or place.

So far, research has focused on the proteins that turn genes on and off in bacteria or in viruses that infect bacteria. In 1981, for instance, Wayne F. Anderson of the University of Alberta, Brian Matthews of the University of Oregon, and their associates deciphered the 3-D structure of a protein called the *cro* repressor at high resolution. The *cro* repressor somehow recognizes a tiny stretch of DNA in a bacterial virus called lambda, latches on to it, and prevents the production of proteins that are important to this virus. Once the structure of the *cro* repressor was in hand, the researchers noted that it has two protruding alpha helices and

A schematic drawing of one subunit of the restriction enzyme *Eco*RI bound to DNA. High-resolution X-ray crystallography of the *Eco*RI-DNA complex showed that a protein can distort DNA while binding to it.

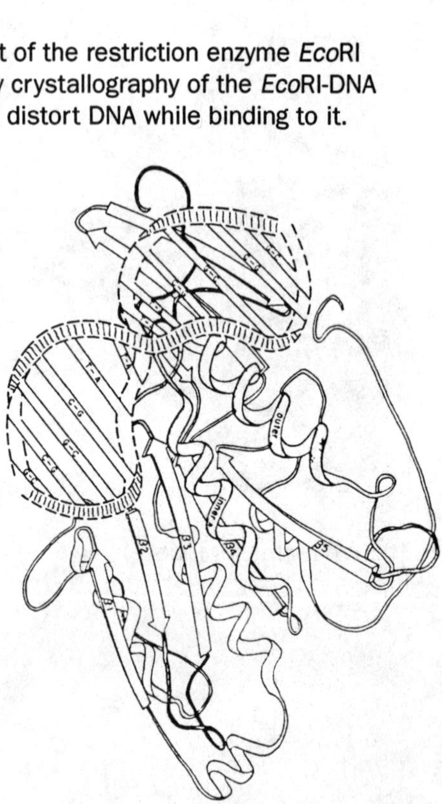

that the space between these helices exactly matches the distance between the major grooves on the surface of the DNA molecule, providing an excellent fit between the two surfaces (see opposite page). This led them to propose a model for the basic mechanism of DNA binding: the helix-turn-helix motif.

Since then, scientists have determined the structures of several other regulatory proteins that have this motif, which consists of an alpha helix, a turn, and another alpha helix. For instance, when Paul Sigler of Yale University deciphered the structure of the repressor that regulates the amount of the amino acid tryptophan in bacterial cells, he found a helix-turn-helix motif.

Having access to these structures spurred researchers to make models of what they thought the structure of the complementary stretch of DNA might be and to use computer graphics to deduce the likely structure of the DNA-protein complex.

Going far beyond such predictions, John Rosenberg of the University of Pittsburgh and his associates actually grew a "cocrystal" that included both a nonregulatory protein (the enzyme *Eco*RI) and its bound DNA, and then solved the structure of this complex at atomic resolution in 1986. Their achievement has been called "a milestone in modern molecular biology." Some scientists found it particularly fitting that it involved *Eco*RI, one of the first of the restriction enzymes that allowed researchers to cut DNA at specific sequences and helped make recombinant DNA technology possible. Without this technology, crystallographers would have been unable to purify sufficiently large samples to grow such cocrystals for analysis.

The high-resolution *Eco*RI/DNA structure proved, for the first time, that a protein can distort DNA while binding to it. Rosenberg calls such distortions in the DNA "neokinks." He defines them as abrupt, localized bends or twists with long-range structural effects on the DNA molecule.

They are not found in DNA that stands alone, but result from the binding of protein to DNA and exist only transiently.

Rosenberg also found that the *Eco*RI does not share the helix-turn-helix motif of the regulatory DNA-binding proteins. Instead, it looks much like the nucleotide-binding domain originally discovered by Rossmann. In addition, it has two "arms" that wrap around the DNA, serving as a sort of anchor and making contact with the major groove in the DNA molecule.

The best way to find out how DNA-binding molecules function is "to get a picture of one of these proteins caught in the act of attaching to its target portion of DNA. That's what we've got now," Rosenberg says. "We can now see certain combinations of structures from the protein and the DNA up against each other in what are some very intriguing patterns. These hold out hope of giving us some general principles of how all these switching mechanisms work.

"Many diseases, including cancer, may be caused by genes which are switched on or off at the wrong time. If we are lucky, this image could provide a clue for how to deal with these diseases in a more effective way."

Studies of other cocrystals in which proteins are bound to very short stretches of DNA are in progress in the laboratories of Stephen Harrison and Mark Ptashne of Harvard University and Carl Pabo of MIT. Pabo has recently turned his attention to determining the structure of a protein-nucleic acid complex that is important in the control of gene expression in higher organisms. He is also exploring a class of proteins that contain domains called "zinc fingers."

Zinc fingers bind to zinc or other metals, as well as to DNA or RNA, and appear to represent another binding motif (in addition to the helix-turn-helix, the mononucleotide-binding site, and others).

Using NMR, several investigators, including Peter Wright of the Scripps Clinic and Research Foundation, have determined the structure of one type of zinc finger. These studies have also provided important information about the biological requirements for DNA binding by zinc fingers.

Zinc fingers and related metal-binding motifs may be involved in the action of several hormone receptors that play important roles in disease. Judging from their sequence of amino acids, for instance, estrogen receptors—which have been linked to breast cancer and are the target of some anticancer drugs—contain a metal-binding domain that may be involved in turning DNA on or off. Pabo hopes that when the structure of these receptors is solved it will lead to better treatments for breast cancer.

SEEKING A CURE FOR THE COMMON COLD

Most of us suffer from at least one cold or other respiratory infection every year. Yet these colds do not give us immunity against additional colds, as would happen in the case of measles or other childhood diseases.

The main reason for our vulnerability is that one of the classes of viruses that cause the common cold—the rhinoviruses—comes in at least 100 varieties, and antibodies against one strain of rhinovirus generally do not offer any protection against the rest.

For the same reason, no one has yet developed an effective vaccine against the common cold. When many different viruses are mixed into a single vaccine, they seriously dilute the vaccine's effect on any one virus.

If one cannot *prevent* colds, can one at least find ways to treat them? This was nearly impossible as long as the structure of the cold virus remained unknown. But now, for the first time, there is real reason to hope for a cure for the common cold. It is just one of the many rewards that can be expected from research in structural biology.

In 1985, using the most advanced techniques of X-ray crystallography, Michael Rossmann and his associates at Purdue University succeeded in solving the structure of human rhinovirus number 14—the first animal virus to be seen at atomic resolution. The structure contained tantalizing clues as to how such viruses infect people, as well as what might be done about it.

The virus consists of a coat of proteins surrounding a strand of RNA. There are four different kinds of proteins in the coat, each one repeated 60 times. Three of these proteins form the coat's outer surface, while the fourth remains on the inner surface. To Rossmann's surprise, the structure of these three outer proteins is extraordinarily similar to the structure of the proteins of several plant viruses.

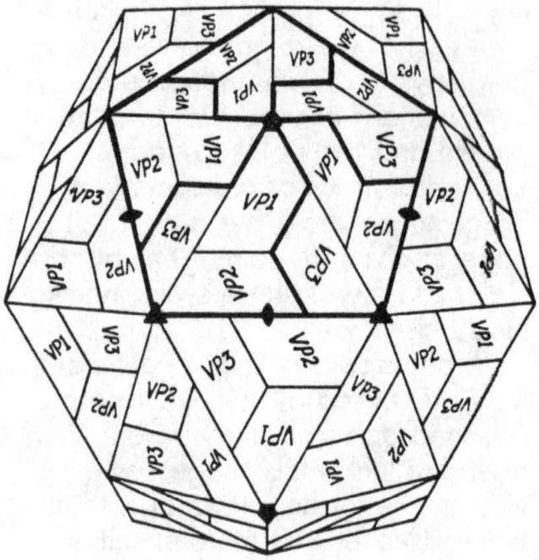

Sketch of the protein coat of rhinovirus 14, showing the three outer proteins that are repeated 60 times.

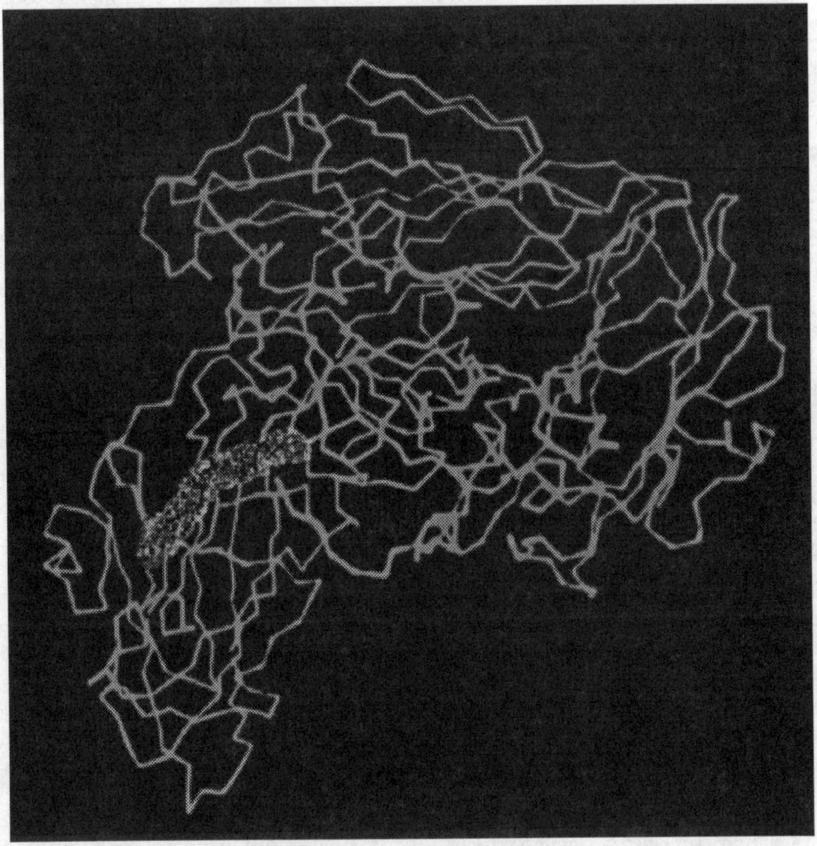

Computer graphic representation of human rhinovirus 14 bound to the antiviral agent WIN 52084.

"These viruses clearly have a common origin," Rossmann says, "and the fact that this structure survived through evolution shows that it is very important to the virus."

Rossmann also discovered a deep cleft, or "canyon," on the surface of the virus, repeated 60 times, which aroused his curiosity. On the rim of the canyon are sites where antibody molecules have been shown to attach themselves to the virus and neutralize it, but the canyon itself is too narrow to let the large antibody molecules in.

"It's a very deep canyon, and the floor of it does not seem to be accessible to antibodies," says Rossmann. "So we postulated that its deeper recesses may be the site at which the virus attaches itself to a cell and attacks it. Only certain kinds of protrusions on the cell would fit all the way into this canyon; that would be the basis of the virus' specificity —why it attacks certain kinds of cells and not others.

"That's what the structure was telling us! And we are slowly finding solid evidence to support this hypothesis."

If Rossmann is right, the segments of viral protein on the canyon's rim that bind to antibodies are likely to mutate to evade these antibodies, thus staying one step ahead of the immune system. But the canyon floor itself "would not be under immune selection and could remain constant, permitting the virus to retain its ability to seek out the same cell receptor," he explains. And in fact, the RNA sequences that code for the proposed antibody-binding sites vary among different viral strains, while the canyon floor is formed by sequences that have been largely conserved from one strain to another.

This system ensures the survival of the virus, but it could also be used by scientists as an elegant target for the design of new drugs.

Recently, a variety of experimental antirhinoviral drugs have been put into crystals of rhinovirus 14 and have been shown to bind to the virus at a specific site. "This is the first time that anyone has been able to see a drug bind to a virus," Rossmann says. "The drugs go to the interior of the viral protein, into the canyon and through a hole, like a trap door, on the canyon floor. They are much smaller than antibodies, so the canyon is no problem for them. It's like putting mules into the Grand Canyon—they can fit in quite well!"

The hole at the bottom of the canyon opens onto a large pocket. When drugs get into this pocket, they interfere with vital stages of the virus' life cycle. The drugs work by altering the shape of the canyon, which in turn alters the ability of the virus to attach to a cell, or by preventing the removal of the viral protein coat after the virus has entered a cell. In either case, the viral RNA is blocked from taking over the machinery of the cell. Viruses cannot reproduce on their own, so they propagate by seizing control of a target cell and forcing it to work for them. After exposure to these drugs, however, the virus' genes are stopped from entering the cell.

These drugs were synthesized by the Sterling Research Group in Rensselaer, New York. When Mark McKinlay and his colleagues screened the drugs for antiviral activity, they discovered that the drugs had an inhibitory effect on the growth of the common cold virus. The structures of the drugs were then modified systematically by other scientists over a period of years until the researchers obtained more powerful compounds with a wider spectrum of action. The Sterling scientists had shown that these compounds prevented the removal of the viral protein coat, but they did not know how the compounds accomplished this until Rossmann and his collaborators did their crystallographic study.

The most exciting feature of these drugs is that they appear to be active not only against the common cold viruses but, to varying degrees, against other members of the "picornavirus" family. These include Coxsackie viruses and the potentially lethal polioviruses. "It's not just a matter of curing the common cold, but of treating other viral diseases," says Rossmann. "Antiviral agents are still very new to us."

The drugs interact with many types of common cold viruses, not just rhinovirus 14. "That's the advantage of these drugs over vaccines," he points out. "Vaccines can act on only one serotype [strain]." In the case of polio, which is caused by three different strains of virus, the widely used vaccine is really a combination of three different vaccines.

The same month that Rossmann published his structure of the common cold virus, James M. Hogle and David Filman of the Scripps Clinic and Research Foundation, together with Marie Chow of MIT, reported the structure of the poliovirus at atomic resolution. It, too, consists of a protein coat (made up of 60 copies of four proteins) around a strand of RNA.

Now that these structures are known, researchers can begin to tailor new drugs that should bind strongly to them. Some structural improvements have already been suggested for the Sterling compounds. Meanwhile, these compounds have been tried on mice and are active at nontoxic oral doses. Human trials of these compounds are now in progress.

As one scientist puts it facetiously, our new ability to see the common cold virus in atomic detail and also to see what various drugs are doing to it "is certainly nothing to sneeze at."

A NEW ERA OF RATIONAL DRUG DESIGN

Until now, the standard method of developing new therapeutic drugs has been to rely on trial and error—or, in some cases, pure luck. But recent developments in structural biology have led scientists to see the possibility of designing drugs as well as various kinds of proteins "rationally"—of creating exactly the right molecule for a specific purpose.

As a result, excitement is gripping researchers who believe that they are entering a golden age of drug design. They foresee new and highly effective treatments against cancer, AIDS, cardiovascular disease, and other currently incurable ills. They expect to produce a new generation of powerful and safe vaccines. They hope to foil bacterial drug resistance.

Their optimism is based on several key advances. First, there is the newly acquired ability to substitute, delete, or insert individual amino acids in existing proteins through a technique called "site-directed mutagenesis," a logical development of recombinant DNA technology. Site-directed mutagenesis is the process of making a deliberate mutation (change) in the DNA sequence that codes for a protein, in order to produce a change at a particular site in the protein. It involves determining the protein's amino acid sequence; translating this into the corresponding DNA sequence; making the desired change at a specific point in the DNA sequence; and putting the mutant DNA sequence into bacteria, which then produce the altered protein.

Scientists have also learned to produce proteins whose amino acid sequence is known by making polypeptides from laboratory chemicals. At the same time, they have a better understanding of protein structures, and particularly of their active sites, thanks to high-resolution X-ray crystallographic studies. NMR is also providing important information, particularly on drug-DNA interactions.

Additional help comes from the growing contribution of interactive computer graphics and the availability of powerful computers for complex calculations. Using these techniques, Irwin Kuntz, Jr., of the University of California, San Francisco, developed a computer program that uses the three-dimensional shape of a molecule, instead of its chemical structure, to identify compounds that might bind to it. This work provides a highly promising method that can be used to search for new drugs. It has already enabled Kuntz to locate a compound that blocks the action of an enzyme essential to the survival of the virus that causes AIDS. Unfortunately, this compound is not effective against the enzyme unless it is used in doses that greatly exceed the lethal limit for human beings. Thus, the compound can only be considered a starting point. Kuntz and his colleagues are now

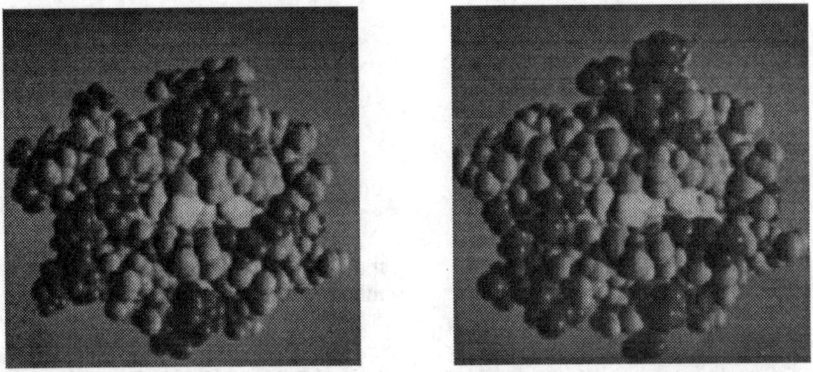

a b

How the native shape (a) of the tuna cytochrome c protein changes (b) when one type of amino acid is replaced by another. In these computer graphic models, lysine (dark, positively charged) is replaced by glutamic acid (light, negatively charged), simulating how protein engineers might reverse the protein's surface charge.

making changes in the compound's structure that they hope will preserve the desired antiviral activity while lessening toxicity.

Finally, there is the recent discovery that antibodies, with their amazing specificity and almost unlimited diversity, can be engineered to work like enzymes and do chemical jobs that could not be performed by any substance before.

These advances open a staggering vista to researchers, who are working to make their vision of radically new, rationally designed drugs come true. But the techniques involved are still in their infancy, and so far the results have been modest.

In 1982, Greg Winter, Alan R. Fersht, and their associates at the Imperial College of Science and Technology in London, England, reported success in deliberately altering a single amino acid of a protein whose crystal structure was known at high resolution. This allowed them to see the results of their site-directed mutagenesis quite clearly in a crystal of the altered protein. Similar studies by other scientists followed rapidly.

Encouraged by this progress, Kevin M. Ulmer, who was then at the Genex Corporation in Gaithersburg, Maryland, wrote a paper outlining the future of what he called ''protein engineering.'' The new technology offers the prospect of creating ''novel proteins not found in nature,'' he wrote. ''By starting with the known crystal structure for a protein, we would like to directly modify the gene to alter that structure in a predictable fashion, targeted to improve some functional property. At each stage we could verify the structural and functional changes that actually occurred and thereby refine and extend our predictive capability.''

This technique was quickly applied to many different proteins whose structures were known. The proteins were altered one amino acid at a time, mostly to learn about their function—for example, to confirm that a particular amino acid was essential for the protein's activity. Going further, some scientists tried to increase a particular protein's stability or its resistance to heat. In a few cases they succeeded, but in others the mutations were clearly harmful.

A surprising number of changes in structure turned out to be "silent," with no obvious effect on the protein's function. However, in certain areas of the protein the tiniest difference in structure had enormous consequences. Some of these critical differences were so subtle that they could not be seen by the standard crystallographic method, which involves placing the images of closely related structures on top of each other and comparing them visually. To find these minute differences, the actual distances between corresponding pairs of atoms had to be calculated.

This is the kind of calculation Joseph Kraut and his colleagues at the University of California, San Diego, made in comparing the 3-D structures of animal and bacterial forms of dihydrofolate reductase, an enzyme that is essential for cell growth. Kraut was puzzled by the fact that trimethoprim, an antibiotic, binds to the bacterial form of this enzyme and inhibits it, thereby killing the bacteria, but has no effect on the animal form of the enzyme. Yet the two structures look very much alike.

"We wanted to know why, with such a high degree of resemblance between the structures of the two dihydrofolate reductases, the drug binds 100,000 times more strongly to bacteria," Kraut says. After examining the high-resolution structures of the two enzymes, each bound to the drug, he actually measured the distances between atoms in these structures and found the answer. "The binding pocket in the vertebrate enzyme was a bit wider, so the drug molecule didn't fit as well," he explains. "It couldn't make as good a contact with the walls of the pocket as in the *E. coli* bacterium because of a 1 to 1½ angstrom (which is pretty small) increase in the width of the binding pocket."

Kraut's conclusion is that the same technique he used "retrospectively," to find out what made trimethoprim a selective drug, could be "run forward, to design a drug that would fit a particular molecule." The idea would be to go through "cycles of iteration." In the first cycle, scientists would design a potential drug that might fit the known structure of the target enzyme (possibly by means of computer graphics) and make this potential drug out of laboratory chemicals, by trial and error. The

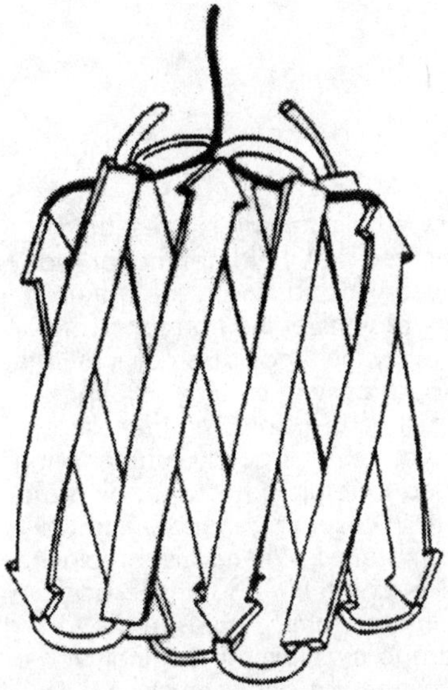

Betabellin, a small protein constructed from laboratory chemicals to help researchers study the folding problem.

scientists would then let the potential drug bind to the enzyme, determine the crystal structure of the enzyme with the potential drug bound to it, and examine the result of this experiment in structural terms. During subsequent cycles they would modify the drug accordingly, and repeat the whole process.

New drugs—from anticancer drugs to antibiotics, antifungal, or antiprotozoan drugs—could be developed much more rapidly in this way, Kraut believes. However, most of these drugs would not be proteins. "We're talking about relatively small molecules that would bind to targets, and the *targets* would be proteins," he explains. "Trimethoprim is only about 20 atoms, a typical small molecule that can fit into an enzyme receptor. Proteins are much larger and more complicated."

Before scientists can effectively design entire proteins with specific properties, he points out, they need to solve the folding problem.

Jane Richardson, the leading classifier of protein shapes, crystallographer David Richardson of Duke University, and Bruce Erickson of The Rockefeller University have embarked on a radical attempt to solve this problem. Instead of taking a natural protein and modifying it, they decided to start from scratch and invented a protein that does not exist in nature.

"A major aim of this project is to learn more about the principles of protein folding," the Richardsons wrote. "We hope, therefore, to make useful mistakes and to learn from them."

Their first creation, which they called "betabellin," is a small protein of 62 amino acids. They designed it to have every possible convenience for researchers: the protein is easy to synthesize, easy to modify, and easy to analyze by means of 2-D NMR as well as by X-ray crystallography (they even provided a built-in heavy-atom derivative to make the crystallography clearer). They expected that it would fold into a barrel made up of two identical beta sheets.

More recently, the Richardsons designed a 79-amino acid protein that folds into a four-helix bundle. They note that this protein, which they named Felix (for "four helix") "is not homologous to any known protein sequence, but is 'native-like' in that it is nonrepetitive and contains 19 of the 20 naturally occurring amino acids." The scientists conclude that their "highly encouraging results" demonstrate the feasibility of this approach to protein design.

Using a different approach, the late Emil Thomas Kaiser of The Rockefeller University focused on the key elements of a protein's structure, rather than on the individual amino acids that make it up. He noted that proteins with similar functions generally have similar structures, even when their amino acid sequences are quite different. His research showed that it did not matter too much which amino acids are used in key regions of the hormone calcitonin, which regulates calcium in the blood, as long as a hydrophilic structure is maintained in these regions. Other parts of the hormone need to retain a hydrophobic structure in order to penetrate into cell membranes. When Kaiser replaced 11 of calcitonin's 32 amino acids with other amino acids, while taking care to maintain the original type of structure, the hormone still did its job.

Next, Kaiser tried to increase the natural hormone's activity by changing only the amino acids that were hydrophilic, leaving the others untouched. His artificial calcitonin turned out to be 2½ times more active in blood than the natural one. In Kaiser's view, the most important aspect of this experiment—beyond the fact that it may lead to better therapy against Paget's disease and other calcium-based ills—was that it represented a step toward learning how to build synthetic active sites and whole enzymes.

Meanwhile, a different group of scientists has launched an ambitious effort to create new "magic bullets" against disease by combining the functions of two separate kinds of proteins, antibodies and enzymes.

Antibodies are produced and sent out by the immune system to fight bacteria, viruses, and other intruders.

They are infinitely varied and choosy, seeking out and binding only to particular types of molecules. But they normally do not speed up chemical reactions as enzymes do. For years, biochemists wondered whether nature's magnificent ability to make a specific antibody for every need could be combined with the chemical power of enzymes to cut through, alter, or break down molecules. Several teams of researchers have now shown that this can be done.

Their method neatly sidesteps the problem of protein folding. They simply let nature make the antibody as usual, but design the antigen, the "foreign" molecule that stimulates cells to produce the specific antibody, so that it resembles an intermediate stage in an enzymatic reaction. The antibody then binds as if it were an enzyme, and the result is an "abzyme"—a protein that has properties of both antibodies and enzymes.

One of the teams involved is headed by Alfonso Tramontano of the Scripps Clinic and Research Foundation; another is led by Donald Hilvert, who is also at Scripps. A third team is headed by Peter Schultz of the University of California, Berkeley. Their work—still in the experimental stage—opens up the possibility that eventually some abzymes may be tailor-made to cut up blood clots and eliminate blockages that could lead to heart attacks or strokes. Others could be made to chew through the protein coats of viruses and thus halt infections. Another group of abzymes might search for, and destroy, cancer cells throughout the body.

Besides, if abzymes can be designed to cut proteins at specific places, they will be extraordinarily useful to researchers. As Schultz points out, they may do for proteins what restriction enzymes have done for DNA: "give us the ability to cut and paste proteins, which could be an enormously important tool for protein study."

NEW TECHNOLOGY WILL SPEED RESEARCH ON STRUCTURES

Structural biology is picking up speed, driven by new developments in technology.

It used to take scientists 5 to 10 years to solve the structure of a protein by X-ray crystallography. One problem was that the diffraction patterns had to be recorded on photographic film, read optically, and then converted to a digital form to determine the intensity of each beam. Alternatively, the patterns could be recorded directly with a counter, but only one beam at a time.

Now, an electronic advance has led to new "area detectors" that combine the advantages of both methods. These new detectors are counters that can distinguish individual beams over a wide area, allowing scientists to record the intensity from all beams at once. As a result, patterns that took months to collect can now be obtained in hours, and with greater accuracy.

For scientists who want to explore how various parts of a protein move and change shape during a chemical reaction, the most exciting development is a breakthrough that allows them to use, in a new way, the high-powered X-ray beams emitted by synchrotrons.

Synchrotrons are massive electron storage rings with which physicists probe the fundamental components of matter. Since their X-ray beams are thousands of times more intense than conventional X-ray beams, all the data that scientists need can be collected in a fraction of the usual time. When these beams are used together with a new X-ray diffraction method based on multiple wavelengths, the necessary exposure time is cut down to milliseconds.

This opens up a possibility that until now researchers had only dreamed of—the possibility of taking a series of X-ray snapshots of a protein structure every few milliseconds. Such a series would allow scientists to track the changes that occur inside the protein during a chemical reaction and actually see some of the short-lived intermediate states for the first time. "Kinetic crystallography" of this sort depends on developing a new generation of area detectors that work rapidly enough to keep up with the changing states. As soon as these become available, researchers expect to make many interesting discoveries.

Other ways to capture what goes on inside a protein—either during its original folding or during a particular chemical reaction—are also being developed. For example, wet specimens can now be frozen so rapidly that the water does not have time to crystallize and the specimen is caught in action. When these samples are examined under an electron microscope, a mathematical analysis of the image can provide three-dimensional details.

Some scientists are studying how proteins fold by combining 2-D NMR with a kind of atomic archeology.

Area detectors such as this NIH instrument allow researchers to speed up the collection of X-ray crystallo-graphic information from months to

Drawing of the synchrotron at Cornell University, Ithaca, New York. By using the powerful X-ray beams from synchrotrons, researchers may be able to track the short-lived motions of protein structures.

They tag the parts of the protein that fold first to distinguish them from those that fold later, so that the entire folding process can be analyzed after it is completed.

Because of all these technical advances, structural biologists appear to be closing in on the folding problem. At the same time, they are beginning to unravel the structures of larger and more complex functional units in human cells—for example, nucleosomes, the structural units of DNA and protein in the nucleus, and ribosomes, the organelles that manufacture proteins inside the cell.

Nucleosomes are now being studied with great intensity because their structure may solve an enduring puzzle: How does a 6-foot length of DNA, which is extraordinarily compacted and tightly supercoiled inside the nucleus, unwind enough to let specific sequences of nucleotides be transcribed into messenger RNA so that they can be expressed as proteins?

Ribosomes have intrigued scientists for years. They are much too big to be analyzed by X-ray diffraction using current methods, yet too small to be seen clearly under the electron microscope. Using a technique called neutron scattering, Peter Moore, Donald Engelman, and their associates at Yale University have mapped the positions of the 21 proteins in the so-called "small" subunit of the *E. coli* ribosome, the subunit that binds to messenger RNA at the beginning of protein synthesis.

An interesting feature of this subunit was demonstrated by Masayasu Nomura, who was then working at the University of Wisconsin. He found that, under certain conditions, the subunit will spontaneously assemble itself from its component parts. "The process is remarkable, and even though we have seen it many times it has not ceased to amaze us," reported Engelman and Moore. "The spontaneous reconstitution of ribosomal subunits implies that the genetic code also contains the information needed to assemble a large complex of protein molecules and RNA."

The study of ever-larger structures—how they are assembled, how they function, and how they affect one another—will lead, inevitably, to a better understanding of the many systems of control in the human body. Structural biology will meet physiology. But this will require larger databases and even greater reliance on the use of computers.

As we decipher these control systems, we will learn how to switch specific genes on or off at will. We will develop a whole arsenal of new drugs, and many new treatments.

Beyond these practical payoffs, structural biology will illuminate some of the most mysterious processes of life. It will help to explain, for instance, how a fertilized egg becomes a human being. And by showing exactly how various proteins interact with DNA, it will bring some answers, at last, to the age-old question of how the environment interacts with our genetic inheritance, in health and in disease.

INSIDE THE CELL

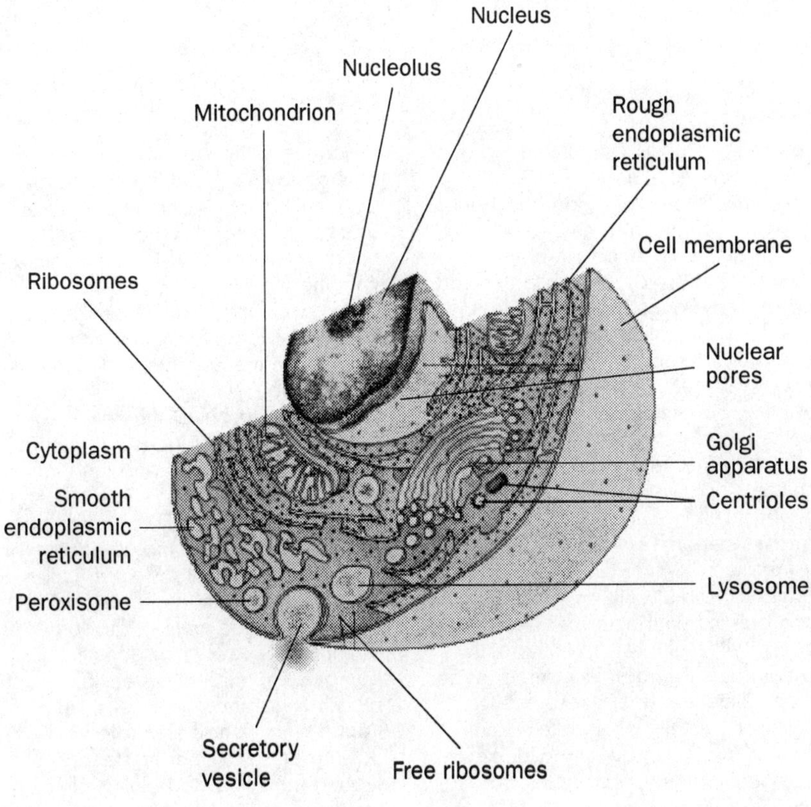

This drawing of an idealized animal cell is based on photographs taken with powerful electron microscopes. Within the cell's membrane are such organelles as the mitochondria (energy producers), the rough endoplasmic reticulum (a site of protein production), the Golgi apparatus (a protein sorter), and the largest organelle, the nucleus (which contains the hereditary material DNA). In addition to these organelles, cells also contain an elaborate network of protein filaments called the cytoskeleton (not shown here) that anchor the organelles, maintain the cell's shape, and direct intracellular traffic.

THE NEXT BIG LEAP

The next big leap in medicine may occur as researchers increasingly apply what they have learned about the world of the living cell to the problems of disease. Although many of the features of cells have been scrutinized and are now quite familiar to students of biology, scientists still want to know more about the architecture of cells, as well as how cells move and grow, how they communicate with one another, and how they work—or fail to work. As scientists study cells in finer and finer detail, the relationship between cell malfunction and certain diseases becomes clearer. Cancer, for instance, is a derangement of cell growth. Normal cells stop dividing upon contact with other cells, but cancerous cells multiply endlessly, as if their surface membranes can no longer perceive signals from nearby cells. If the still-cryptic signals that tell a cell when to divide and when to stop are deciphered, new approaches to cancer treatment could result. Other major health problems of today, including genetic disorders, heart disease, and diabetes, may also yield to improved therapies as more is learned about the cell.

The work is already under way. In the past 40 years, through a combination of electron microscopy, biochemical analyses of cell parts, X-ray crystallography, and other techniques, cell biologists have looked inside the cell and revealed it to be a complex and highly organized entity. A typical cell is like a miniature body containing tiny organs, called organelles. One organelle is the command center, others provide the cell with energy, while still others manufacture proteins and other molecules that the cell needs to survive and to communicate with the world around it. The entire cell is enclosed in a fine skin, its surface membrane. This membrane not only keeps the cell intact, it also provides portals that allow selected molecules into and out of the cell.

This booklet describes the internal landscape of the cell and the work of some of the pioneers who first mapped its features. Much has been glimpsed, but much more remains to be seen. How, for instance, is the work of protein manufacture orchestrated within the cell? What prompts a cell to divide? How does a human being, complete with eye, hair, bone, blood, skin, and nerve cells, arise from a single undifferentiated cell? Solving these cellular mysteries will require the concerted efforts of scientists from many disciplines, including anatomy, physiology, physics, and chemistry. Cell biologists of today can look forward to a future of discovery and refined understanding as they continue to reveal the living worlds within us.

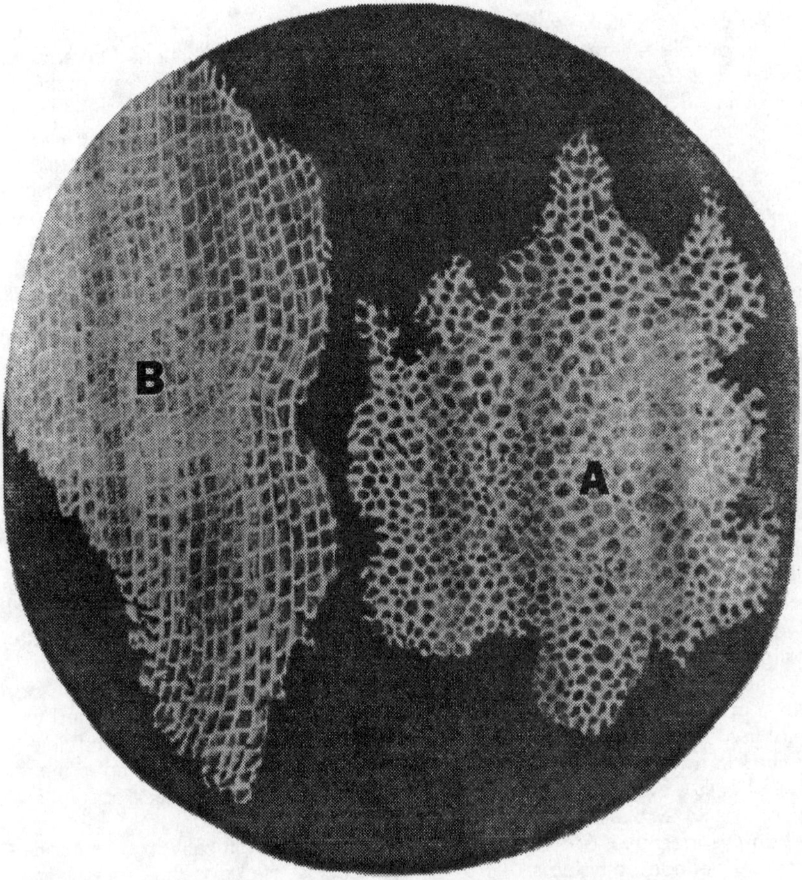

This drawing of cork tissue, as seen under a simple microscope, appeared in Robert Hooke's 1667 book, *Microscopy*. Hooke named the compartments "cells."

WHAT ARE CELLS?

The idea that every living thing is made up of cells emerged from an encounter between two German scientists in 1838. Nearly two centuries earlier, in 1665, the English physicist Robert Hooke looked at a sliver of cork through a microscope lens and noticed some "pores" or "cells" in it. Hooke believed the cells had served as containers for the "noble juices" or "fibrous threads" of the once-living cork tree. He thought these cells existed only in plants, since he and his scientific contemporaries had observed the structures only in plant material.

During a dinner conversation in 1838, however, the botanist Matthias Schleiden, who had been studying plant cells, and the zoologist Theodor Schwann, who had been examining the nervous tissue of animals, realized that the similarities between the structures they had been investigating were too strong to be accidental. In 1847, Schwann wrote a paper describing how all animal tissue, including bone, blood, skin, muscle, and glands, is composed of cells. Even sperm and eggs are cells. Schleiden elaborated on this idea as it applied to plants. A German pathologist, Rudolph Virchow, is given credit for being the first to state, in 1858, what became known as the cell theory: "Every animal appears as a sum of vital units, each of which bears in itself the complete characteristics of life."

The cell theory united plant and animal sciences by recognizing that the cell is the fundamental component of all living organisms, from orchids and earthworms to human beings. It provided an intellectual framework that revealed the hidden similarities of form and function in extremely diverse organisms, and it gave scientists a way of making sense out of the bewildering array of living creatures. But what is a cell?

Obviously, there are major differences among cell types. Muscle cells, which can contract, have to be quite different from liver or bone cells. Nerve cells have long, thin fibers that, in humans, might extend more than 3 feet from the spinal cord to the toes, while blood cells have no projecting fibers at all. Plant cells have a unique ability to use light for energy.

Then what do all these cells have in common? Discovering their shared properties was difficult. At first, scientists thought that the cell was just a blob of jelly, or some primordial soup enclosed in a bag. They named the jelly "protoplasm." For a long time they could not find anything in the protoplasm, which later became known as the "cytoplasm."

Part of the difficulty in studying cells, of course, is due to their extremely small size. The cells of multicellular organisms are impossible to see with the unaided eye. Schleiden and Schwann, like cell biologists before and after, relied upon microscopes to enlarge the image of cells so that they could be studied. Microscopes employ one or more curved lenses and a

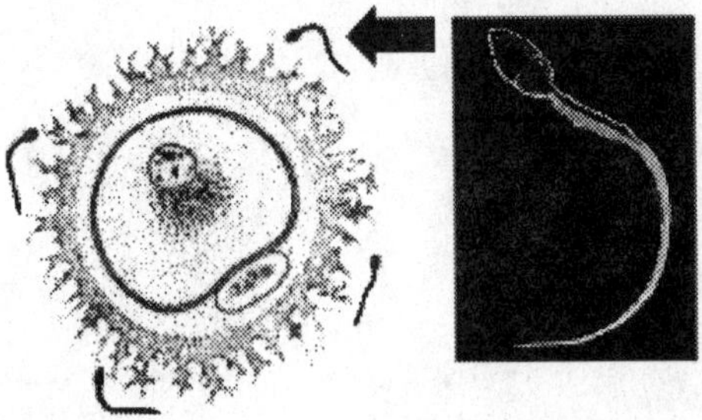

Egg cell Sperm cell

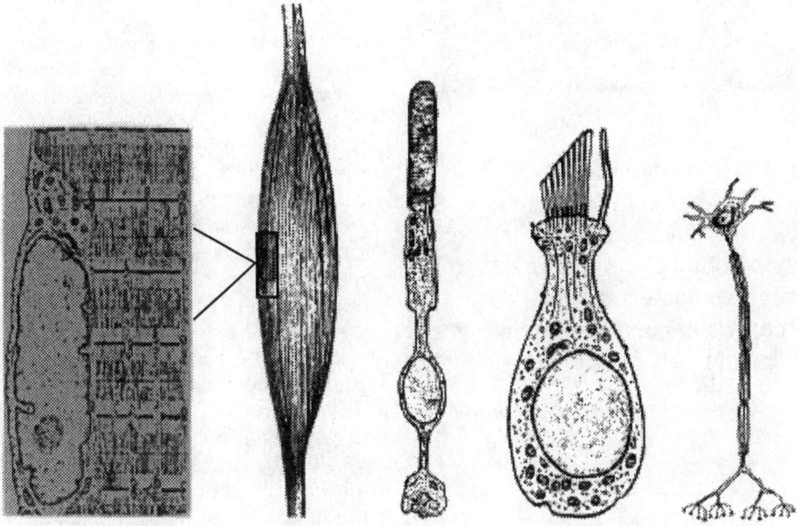

Muscle cell Rod cell in the eye Hair cell Nerve cell

The variety of human cells. After fertilization by a sperm, a single human egg cell divides again and again into many kinds of specialized cells whose structures vary according to the functions they fill. Some nerve cells, for example, are 3 feet long to reach from spine to toe. The orderly structure of typical skeletal muscle is shown here in such detail that if the entire muscle cell were drawn to the same scale as the fragment shown, it might be 1,000 feet long.

Anton van Leeuwenhoek (1632-1723) was an early microscopist who ground his own lenses as a hobby and was the first to observe such living cells as sperm and pond water microorganisms.

source of illumination (typically white light) to magnify cells.

One of the most remarkable early microscopists was a Dutch draper named Anton van Leeuwenhoek, who ground his own lenses as a hobby. Van Leeuwenhoek, who once made a lens from a grain of sand, used simple (single-lensed) microscopes to examine everything from pond water to the scum on his teeth.

In 1702, van Leeuwenhoek reported to the Royal (Scientific) Society of London that he had observed "a little clear sort of light in the middle" of a fish blood cell he had been examining. This description of what was later called the cell's nucleus was the first suggestion that animal cells had an internal structure. Throughout the 18th and 19th centuries, improvements in microscopes and techniques for selectively staining cell parts enabled cell biologists to distinguish other particles within the cell. However, researchers could not study these minute flecks in detail because they met an insuperable obstacle: the wavelength of light.

A light microscope—even one with perfect lenses and perfect illumination—simply cannot be used to distinguish objects that are smaller than one-half the wavelength of light. White light has an average wavelength of 0.55 micrometers, half of which is 0.27 micrometers. (One micrometer is a thousandth of a millimeter, and there are about 25,000 micrometers to an inch. Micrometers are also called microns.) Any two

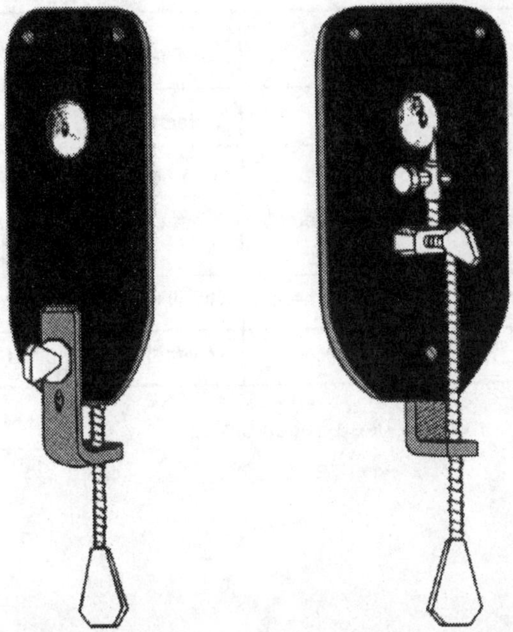

This is the actual size of a typical microscope built by van Leeuwenhoek. He peered through the tiny lens opening on one side of a metal plate (left) to see the specimen mounted on the point of a pin on the other side (right). The specimen could be moved into focus by a system of screws.

lines that are closer together than 0.27 micrometers will be seen as a single line, and any object with a diameter smaller than 0.27 micrometers will be invisible—or, at best, show up as a blur.

Although the nucleus of a typical human cell is relatively large (about 7 micrometers in diameter), most organelles vary from a width of only 1 micrometer to structures so fine that they must be measured in nanometers (which are 1,000 times smaller than micrometers), or even in angstrom units (10 times smaller than nanometers). To see such tiny particles under a microscope, scientists must bypass light altogether and use a different sort of "illumination," one with a shorter wavelength.

The invention of the electron microscope in the 1930's filled the bill. In this kind of microscope, electrons are accelerated in a vacuum until their wavelength is extremely short—only one hundred-thousandth that of white light. Beams of these high-speed electrons are focused on a cell sample and are absorbed or scattered by the cell's parts so as to form

Unit	Equal to	Used to Measure
Centimeter	1/100 meter	Objects visible to the eye
Millimeter	1/10 centimeter	Very large cells
Micrometer (or Micron)	1/1000 millimeter	Most cells, large organelles
Nanometer	1/1000 micrometer	Small organelles, large molecules
Angstrom	1/10 nanometer	Molecules, atoms

Units of size commonly used in cell biology.

an image on an electron-sensitive photographic plate.

If pushed to the limit, electron microscopes can resolve objects as small as the diameter of an atom. Most electron microscopes used to view biological material can "see" down to about 10 angstroms—an incredible feat, for although this does not make atoms visible, it does allow researchers to distinguish individual molecules of biological importance. In effect, it can magnify objects up to 1 million times. Nevertheless, all electron microscopes suffer from a serious drawback. Since no specimen can survive under their high vacuum, they cannot show the ever-changing movements that characterize a living cell.

The first electron microscopes were used to study crystals and were impractical for the study of cell structure. Cell researchers had to learn how to cut extremely thin slices of cells, sometimes down to a thickness of only a few hundred angstroms, so that electrons could pass through them. Also, to ensure contrast between different parts of the otherwise transparent cell, new staining techniques had to be devised, involving "heavy" metals that are absorbed to differing extents by various cell parts. The cell sections also had to be "fixed" in new ways, to preserve them, and embedded in new kinds of materials (mostly transparent plastic). Altogether, it was not until the early 1950's that electron microscopes began to be used routinely. While microscopists were looking through their instruments at smaller and smaller particles in cells and attempting to understand their structure, another group of scientists was pursuing an entirely different, but equally important, line of research—biochemistry.

The size of common objects as viewed under microscopes.

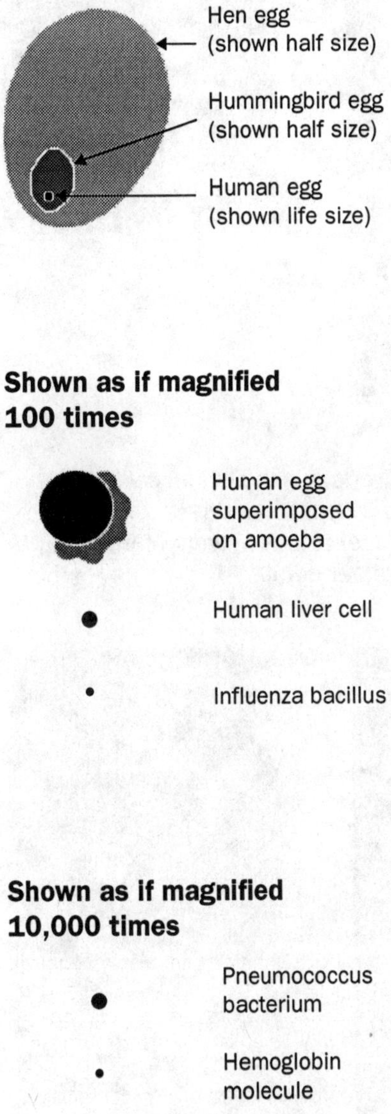

Diagrams of the parts of a light microscope (top) and an electron microscope (bottom).

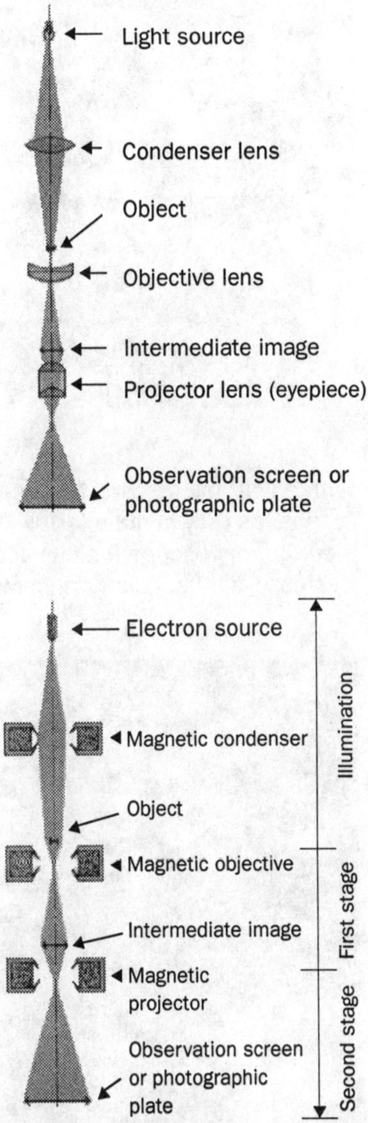

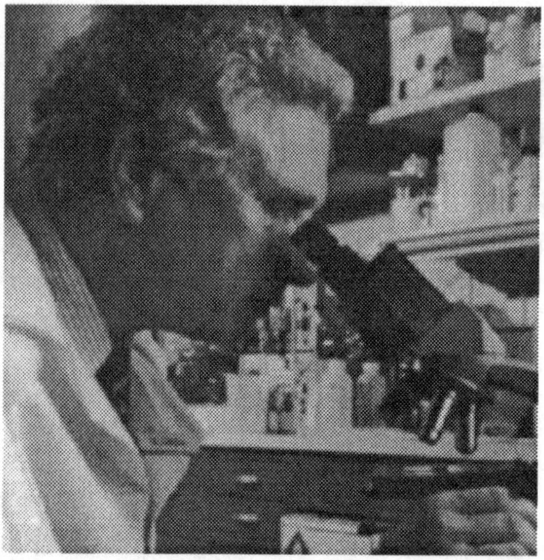

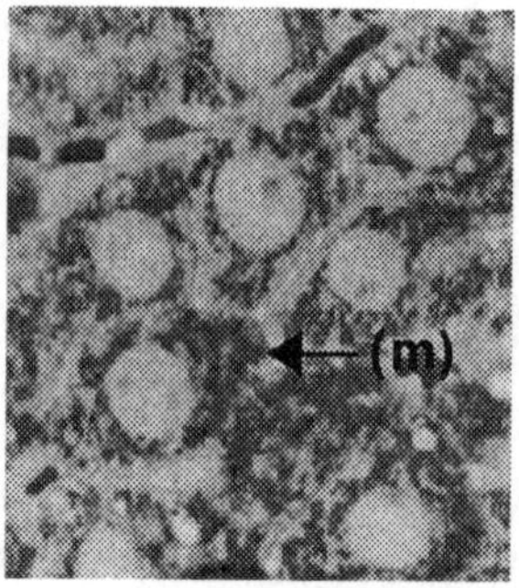

With a light microscope (above), a viewer can see several cells and the organelles they contain (arrow indicates a mitochondrion), while an electron microscope (below) allows a user to see objects of interest (in this case a mitochondrion) in much greater detail.

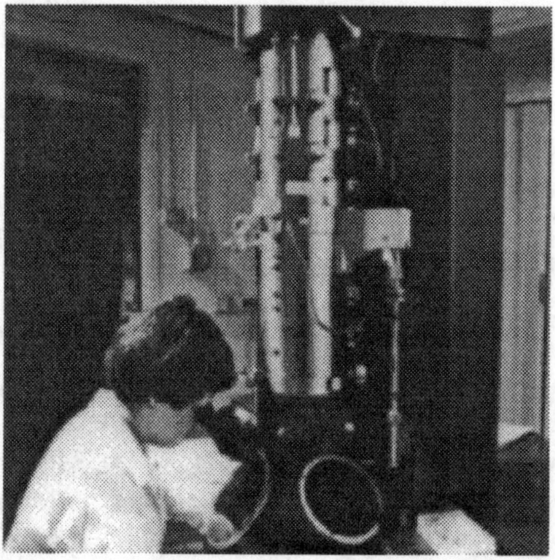

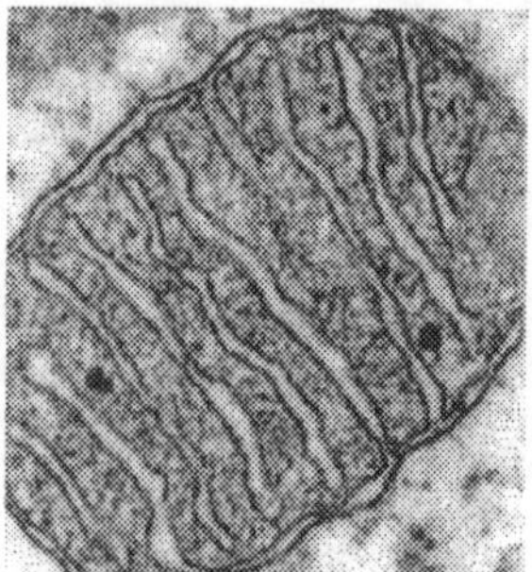

Modern Microscopy Gives Clear View of Cell Structure and Movements

Using an instrument the size of his palm, Anton van Leeuwenhoek was the first person to study the movements of living sperm. Modern descendants of van Leeuwenhoek's light microscope can be over 6 feet tall, but they continue to be indispensable to cell biologists because, unlike electron microscopes, light microscopes enable the user to see living cells in action. The primary challenge for light microscopists since van Leeuwenhoek's time has been to enhance the contrast between pale cells and their paler surroundings so that cell structures and movement can be seen more easily. Recently, ingenious strategies involving video cameras, polarizing light, digitizing computers, and other techniques have yielded vast improvements in contrast and have fueled a renaissance in light microscopy.

A polarizer causes light waves to move in parallel planes, thus reducing the distortion that results when light scatters across a magnified object. This technique was first used in the 1950's, and provided many new clues to cellular activities, particularly the intricacies of cell division. Further enhancements in visualizing the cell's interior came in the early 1980's when microscopists Shinya Inoué of the Marine Biological Laboratory in Woods Hole, Massachusetts, and Robert Allen of Dartmouth College began turning video cameras onto living cells. Unlike the eye, a video camera can "see" objects clearly even when the contrast between subject and background is very poor. Inoué and Allen used video cameras to watch food-containing vesicles and other bodies in the cell move rapidly along slender track-like organelles. Video images can now be further enhanced by digitizing computers, which, when attached to a camera, scan the cell, break down the image into light and dark bits, and then reconstruct the image so that "visual noise" (grayness) is subtracted, while objects of interest are highlighted.

A new type of microscope, called the confocal microscope, promises to have a great impact on the study of cell structure. A confocal microscope passes a beam of light over a tiny portion of a cell, then focuses the light that reflects off the specimen through a pinhole. A sharply focused, three-dimensional image of a cell or cell structure can be built up by recording the intensity of the light beam coming off each scanned point and then reconstructing the whole image on a viewing screen. Because confocal microscopes can be used on living cells, they allow researchers to see clearly cell movements and the interactions of neighboring cells.

Electron microscopy is also undergoing some exciting developments. One new instrument that has aroused great interest is the scanning tunneling electron microscope. This device consists of an ultrathin tube that is held a fraction of a millimeter away from a cell or other sample. The distance between the sample and the tube is reduced until a current of electrons jumps from the sample and travels up the tube, a phenomenon called tunneling. As the tube is moved over the surface of the sample, a computer records the distances between the sample and the tube and converts this information into a picture of the sample's contours. In 1989, researchers reported using scanning tunneling electron microscopy to obtain, for the first time, direct images of pure DNA. Eventually, this method may be used to observe living viruses or the actions of molecules on the cell surface. While these latest revolutions in microscopy are still in their early stages, they are already enabling scientists to see the tiniest details of cell structure and activity in ways undreamed of a few years ago.

A FRUITFUL FUSION OF TWO TECHNIQUES

The study of biochemistry goes back to Antoine Lavoisier, the 18th-century French scientist who explained the role of oxygen in the respiration of both plants and animals, established the composition of water and other compounds, and introduced quantitative methods in the study of chemical reactions, thereby laying the foundation for modern chemistry.

In the 19th century, biochemists isolated and identified many cellular chemicals—for example, hemoglobin, the red pigment of blood, and chlorophyll, the green pigment in plants. They discovered that compounds taken from animal tissue consisted of the same chemical elements as nonliving materials. They isolated the nucleic acids, which are now known to govern heredity and protein synthesis. They began to study proteins, especially enzymes, which catalyze chemical reactions in cells.

When dealing with cells, biochemists behave quite unlike microscopists, who have enormous respect for the details of the cell's structure. Biochemists simply grind up large quantities of cells to release their contents into a solution (this is called "homogenizing") and then analyze the mixture (called the homogenate). Often, the homogenate is fractionated, or separated, into individual components. Usually this is done with a centrifuge, a machine that separates particles according to their size and density by whirling them around at varying speeds. The heaviest and largest particles are thrown to the bottom of the test tube most rapidly, followed by somewhat lighter and smaller components, until at the highest speed there remain only the smallest and lightest particles at the top.

In 1925, a Swede, Theodor Svedberg, developed an instrument that would prove at least as revolutionary as the electron microscope: the ultracentrifuge, a machine that could spin its samples at such high speeds and with such force (it could attain hundreds of thousands of times the force of gravity) that many of the smaller and lighter components of the cell and even proteins and nucleic acids could be collected separately and studied for the first time.

The significance of this new instrument did not become apparent until years later. For a long time, biochemists seemed interested only in the chemical reactions of the cell as a whole—for example, how the cell obtains energy or synthesizes proteins. The scientists gave little thought to what the various fragments they dealt with represented in the cell—what they looked like, how they were organized, or how they related to one another. Very often, biochemists gave these fragments names of their own, unaware that microscopists had already examined and named them.

"There were two classes of people, and they didn't communicate with each other at all!" recalls DeWitt Stetten, Jr., Deputy Director for Science, Emeritus, at the National Institutes of Health (NIH). Finally, in the 1950's, the two groups began to

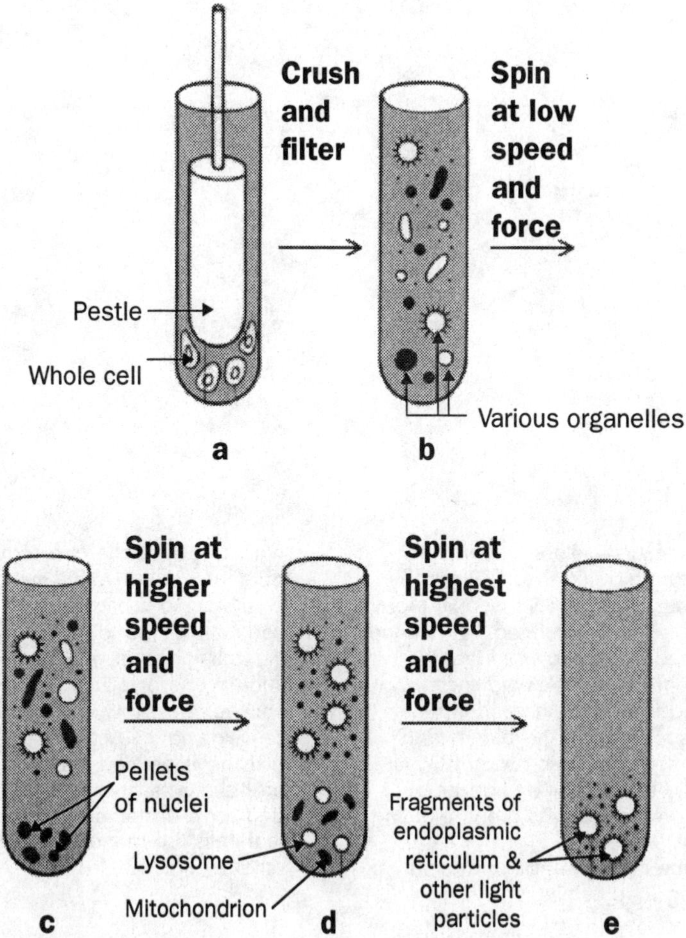

How a centrifuge is used to isolate cell components. To separate the various particles in cells, biochemists begin by placing whole cells in a solution and then breaking the cells with a pestle or with high-frequency sound waves (a). The mixture is then filtered to remove unbroken cells. At this point, the cell organelles and fragments are free-floating (b). As the sample is spun in the centrifuge at increasingly higher speeds and force, the organelles begin to settle to the bottom depending on their size and density. First to settle out are pellets of nuclei (c). At forces greater than 10,000 times that of gravity, pellets of mitochondria and lysosomes sink to the bottom (d). Finally, at very high speeds and at forces 100,000 times that of gravity, the very lightest particles and organelles begin to settle out (e).

edge closer together. "It was an important fusion," says Stetten. By that time, electron-microscopic techniques had been refined. When the microscopists discovered that the biochemists' particles matched what they had been seeing under their microscopes, and the biochemists learned that the microscopists could actually see what they had been analyzing, there was great rejoicing on both sides.

As the microscopists and biochemists began to communicate, there was an avalanche of discoveries about the world within the cells of animals and plants. A whole new vocabulary had to be developed for the cellular structures that the electron microscope and ultracentrifuge uncovered. Now, after 40 years of painstaking investigation, cell biologists have defined many of the general characteristics that cells share, and have discerned the mechanisms that cells use to make enzymes and other vital molecules.

Enzymes — The Catalysts of Life

Without enzymes, there would be no life. Nearly all of the myriad chemical reactions occurring within a cell at any moment require the participation of at least 1 of the 4,000 or so enzymes in the cell's repertoire. Many cellular processes are energetically unfavorable; without enzymes they would proceed slowly or not at all. Enzymes act as catalysts, speeding up reactions without being permanently altered themselves. Thus, enzymes can do their jobs—often, cutting apart or splicing together other molecules—over and over. According to chemist Ronald Breslow of Columbia University, enzymes work so well that a process that takes 5 seconds (such as reading this sentence) with enzymes would take 1,500 years without them. Enzymes also have great specificity; like a lock, each enzyme will accept only appropriately shaped "keys" (called substrates).

With few exceptions, enzymes are proteins and are made of strings of amino acids ordered according to instructions contained in the genes. Even if the sequence of amino acids in an enzyme is known, however, scientists cannot predict how the enzyme will fold into its final, active shape. This is the so-called "folding problem" that many researchers are working to solve. If scientists can learn the rules by which enzymes and other proteins fold, it would open the way to synthesizing engineered, artificial enzymes with therapeutic, industrial, and manufacturing applications.

The first exception to the rule that all enzymes are proteins was uncovered in the early 1980's when Thomas Cech of the University of Colorado at Boulder and other scientists made the startling discovery that the nucleic acid RNA can act as an enzyme because it can cut and splice itself. This observation, which led, in 1989, to a Nobel Prize for Cech and Sidney Altman of Yale University, has caused some scientists to speculate that RNA was the first self-forming and self-reproducing molecule to evolve.

If an enzyme is missing or malfunctioning, one of a large class of diseases, collectively called metabolic disorders, may develop. To date, most attempts at treating such diseases in humans with enzyme replacement have failed because the body quickly breaks down ingested or injected enzymes. In 1987, however, researchers at Duke University developed a chemically "camouflaged" enzyme that can escape detection by the antibodies that attack foreign molecules. It has shown promise in treating children who are deficient in adenosine deaminase, an enzyme that plays a crucial role in the immune system.

THE ASTONISHING UNIFORMITY OF LIFE

All cells—whether from a bacterium, plant, mouse, or human—are made of the same basic materials: nucleic acids, proteins, carbohydrates, water, fats, and salts.

"The uniformity of the earth's life, more astonishing than its diversity, is accountable by the high probability that we derived, originally, from a single cell," notes physician Lewis Thomas in *The Lives of a Cell*. "It is from the progeny of this parent cell that we take our looks; we still share genes around, and the resemblance of the enzymes of grasses to those of whales is a family resemblance."

The genetic material in all these cells is deoxyribonucleic acid (DNA), a large molecule that directs the making of duplicate cells. DNA also directs the building of proteins according to a complex code. Even the simplest living cells—the mycoplasma—contain a relatively large amount of DNA, enough to code for up to a thousand different proteins. Every human cell has about 6 feet of very tightly wound DNA strands contained within its nucleus, and every adult carries about 100 billion miles of ultrathin DNA strands in his or her body—a distance greater than the diameter of the solar system.

Each cell is separated from the rest of the world by a membrane so thin that it cannot be seen under a light microscope. Despite its ethereal nature, the surface membrane is exceedingly powerful, controlling everything that goes into and out of the cell and relaying vital messages. Similar membranes enclose or make up a large number of the cell's organelles.

There is a fundamental distinction between the two major categories of cells. Prokaryotic cells, which include bacteria, mycoplasma, and blue-green algae, do not have a membrane around their nuclear region. Eukaryotic ("proper nucleus") cells, in contrast, have a double membrane separating the nucleus from the cytoplasm, as well as many other internal membranes to segregate their organelles. The cells of all animals and plants (except blue-green algae) and one-celled protozoa are eukaryotes.

Only eukaryotic cells are able to combine with one another to form multicellular systems—an important step up the evolutionary ladder. And while, in general, prokaryotic cells produce only exact duplicates of themselves, eukaryotic cells are capable of differentiation into many kinds of cells, at least in higher organisms. This gives eukaryotic cells certain obvious advantages. However, prokaryotes have advantages of their own: simpler nutritional requirements and much more rapid growth and division. As Daniel Mazia of Stanford University points out, the differences between the two types of cells are simply "different ways of making a living." A cell's job—which, for many cells, centers on the manufacture of proteins—is divided into many steps. In eukaryotes, the production of protein begins in the most prominent organelle—the nucleus.

a cell are "on," or "expressed," and giving orders for the production of specific proteins. Some of the instructions that switch these genes on or off come from the cytoplasm, where they are generated as a result of interactions between the surface membrane and the environment. Thus, the commands from the nucleus are influenced by what goes on outside the cell as well as by the cell's genetic program.

An order to make a protein begins when the appropriate genes are "transcribed" from the DNA into strands of another kind of nucleic acid, called messenger ribonucleic acid (mRNA). Messenger RNA is manufactured by transcribing just one chain of the DNA double helix (one side of the twisted ladder). A strand of mRNA is complementary to the DNA from which it is transcribed, except that each adenine of the DNA is paired with a uracil (U), instead of with thymine. For example, a stretch of DNA bases ATCG is transcribed into the mRNA sequence UAGC.

After additional processing, the mRNA carries its message into the cytoplasm, while the DNA remains safely in the nucleus, somewhat like the printing block in a printing press. It may be that the mRNA gets out through "gates" in the nuclear membrane called nuclear pore complexes, although this theory is still unproved.

Once in the cytoplasm, the messenger RNA carries its instructions to tiny organelles called ribosomes, the "factories" in which the next step of protein manufacture, called translation, takes place.

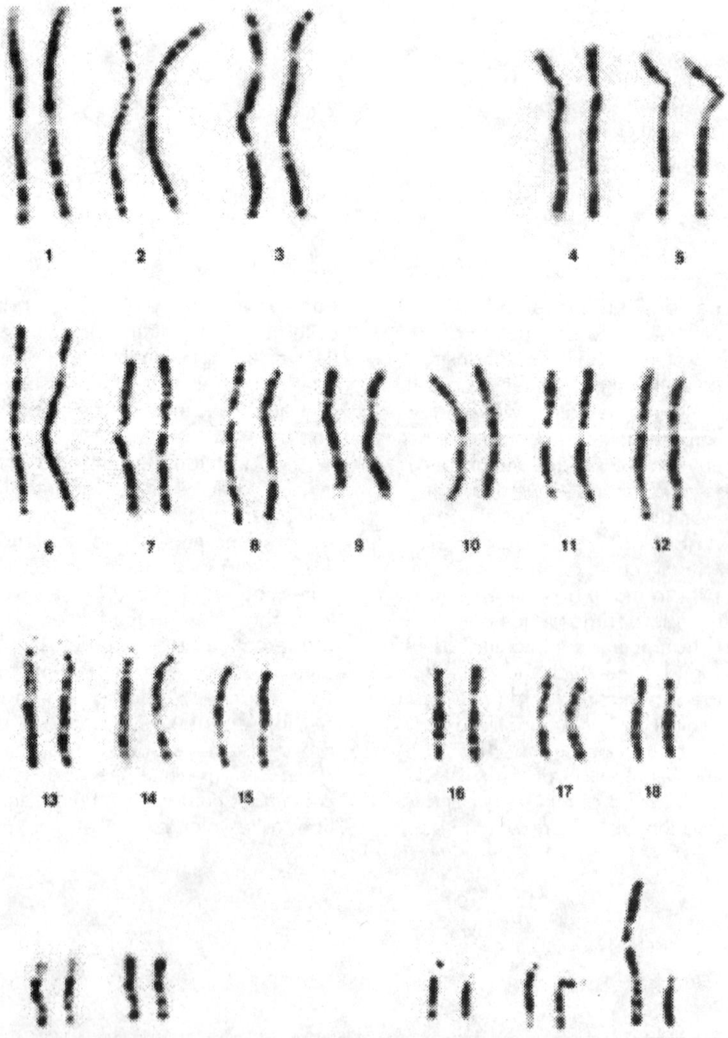

Chromosomes of a normal human male.

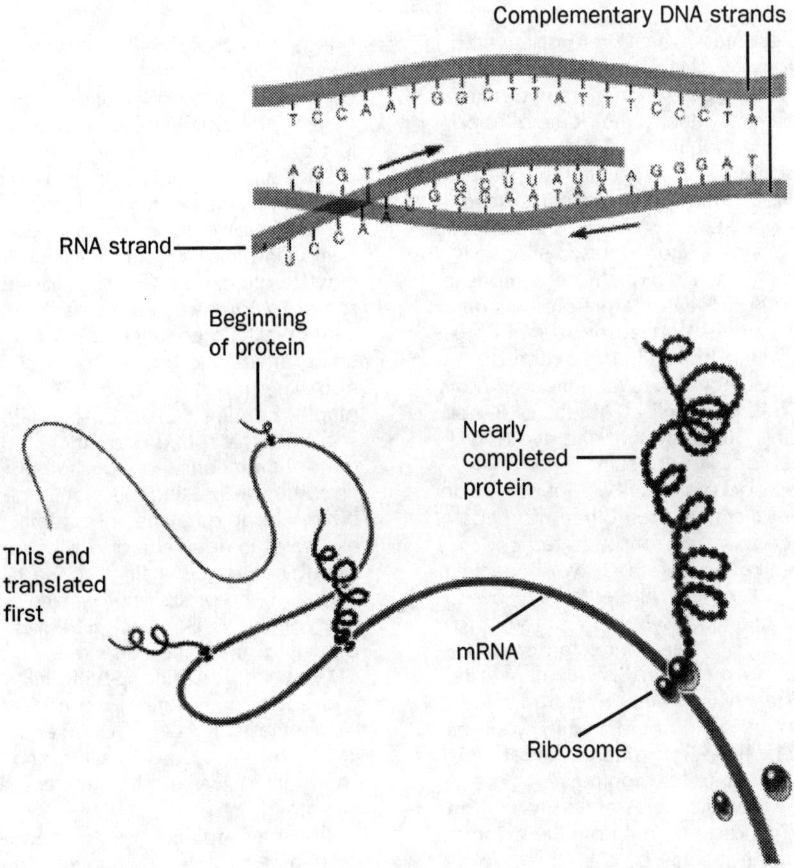

In the nucleus, DNA's instructions are transcribed (top) into a messenger molecule of ribonucleic acid (RNA). The code in a strand of messenger RNA is translated into a protein (bottom) in tiny organelles, called ribosomes, in the cytoplasm.

RIBOSOMES, THE "PROTEIN FACTORIES"

Ribosomes, which were discovered in the mid-1950's, are extremely tiny—less than 30 nanometers in diameter. However, due to their crucial role in protein manufacture, ribosomes can also be extremely numerous. In *E. coli*, for example, ribosomes account for one-fourth of the cell's mass. Each ribosome is made of two unequally sized subunits, which are composed of at least 40 different proteins and a form of RNA called ribosomal RNA.

During translation, a strand of mRNA moves between the two parts of a ribosome like a piece of thread being pulled through the eye of a needle. The ribosome "reads" the message of the mRNA not one nucleotide at a time, but rather in groups of three. These groups, called codons, are like words. Each word specifies one of the 20 different amino acids (the chemical subunits that form proteins) or is a signal to start or stop making a protein. For example, the codon AGC in mRNA is translated into the amino acid serine, whereas nucleotides in a different order, say GCA, code for alanine.

The amino acids called for by the mRNA are brought from the cytoplasm to the ribosome by a third kind of RNA, transfer RNA (tRNA). This small molecule is a connector: one end carries three nucleotides, known as the "anticodon," which will join to a codon in the mRNA according to the rules of base pairing (A with U, and G with C). The molecule's other end carries an amino acid. As the mRNA passes through the ribosome, tRNA brings the correct amino acids in and they are linked together by peptide bonds to form a polypeptide chain. When all the amino acids for a protein are joined, the chain is released.

Each strand of mRNA can be read many thousands of times. Indeed, at any one moment a strand of mRNA containing the transcript for a protein may be attached to 30 separate ribosomes. Moreover, ribosomes work very quickly to connect the required amino acids into a protein. Each ribosome in a single *E. coli*, for example, can link 15 amino acids in a second. The speed and efficiency of translation means that each gene is capable of directing the manufacture of very large quantities of protein. For example, in each cell of a silk worm's silk gland there is a single gene that codes for the protein fibroin, the chief component of silk. Each time it is activated, the gene can make 10,000 copies of its specific mRNA, and *each* copy of mRNA can direct the synthesis of 100,000 molecules of fibroin. In 4 days, a silk gland cell can manufacture a billion molecules of fibroin!

Ribosomes fall into two categories: those that are free in the cytoplasm and those that are bound to membranes. The two kinds of ribosomes play similar roles in the manufacture of proteins. But while the free ribosomes leave the proteins equally free to float in the cytoplasm, the bound ribosomes transfer their finished proteins into a large, cobwebby organelle—the endoplasmic reticulum.

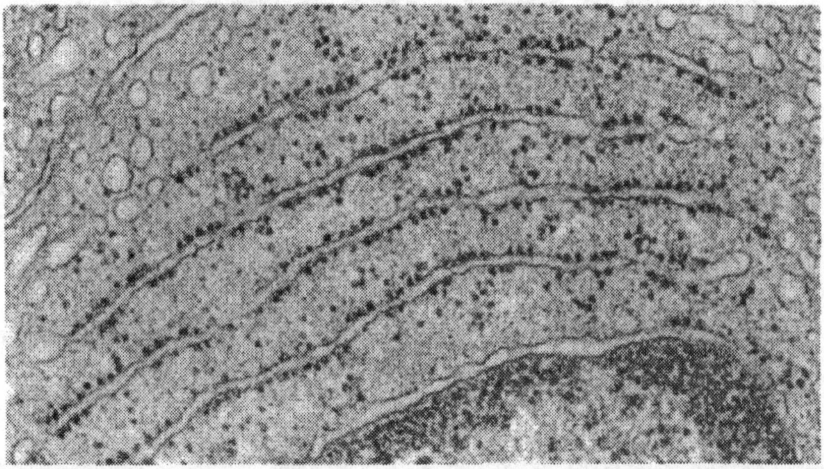

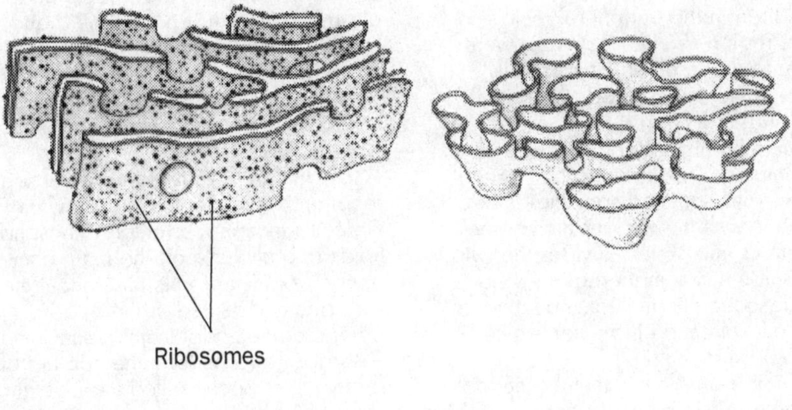

Electron micrograph (top) shows the folds of the endoplasmic reticulum thickly dotted with tiny dark bodies, the ribosomes. Drawings (bottom) show both the ribosome-covered rough endoplasmic reticulum (left) and the ribosome-free smooth endoplasmic reticulum (right).

THE ENDOPLASMIC RETICULUM

In 1945, as the electron microscope was becoming a useful research tool, Albert Claude of Belgium and Keith Porter, who was then at The Rockefeller Institute, used it to discover a vast network of channels bounded by membranes in the cytoplasm of chick embryo cells. At times this network looked like the concentric circles of a slice of onion. Porter called this network the endoplasmic reticulum because it was more concentrated in the inner (endoplasmic) region of the cell than in the peripheral (ectoplasmic) region. Similar networks were later found in all eukaryotic cells, except mammalian red blood cells.

It was discovered that the membranes of this endoplasmic reticulum all interconnect, forming a system of tubes and flattened sacs that is continuous with the nuclear membrane. In effect, this system divides the cytoplasm into two main regions, one enclosed within the "plumbing" and the other forming the outer region, or cytoplasmic matrix.

Some parts of this membrane look smooth, while others appear "rough" because they are dotted with ribosomes that form granules on their outer surfaces. These ribosomes deposit newly formed proteins into the lumen, or inner space, of the endoplasmic reticulum. The endoplasmic reticulum then segregates the proteins into those that will be needed in the cytoplasm and those that will be transported to other organelles or secreted from the cell.

In the mid-1950's, George Palade, then of The Rockefeller Institute, concluded that the amount of rough endoplasmic reticulum in a cell corresponds closely to the quantity of protein the cell exports. Those white blood cells that produce infection-fighting immune system proteins called antibodies have highly developed rough endoplasmic reticula, for example. These antibodies are found mainly in cellular storage areas, from which they go forth to combat infections. The smooth endoplasmic reticulum, on the other hand, is particularly well developed in cells where it takes on some extra function —for example, in liver cells, where it breaks down drugs by making them water soluble.

In addition to its role in protein segregation, the endoplasmic reticulum is the cell's membrane factory. Phospholipids and cholesterol, the main components of membranes throughout the cell, are synthesized in the smooth portion of the endoplasmic reticulum. These compounds form the coating of protein-filled sacs, called vesicles, that "bud off" from the endoplasmic reticulum, migrate to another organelle, fuse with it, and then deposit the protein cargo.

Most of the proteins leaving the endoplasmic reticulum are still not mature; they must undergo further processing in another organelle, the Golgi apparatus, before they are ready to perform their functions within or outside the cell.

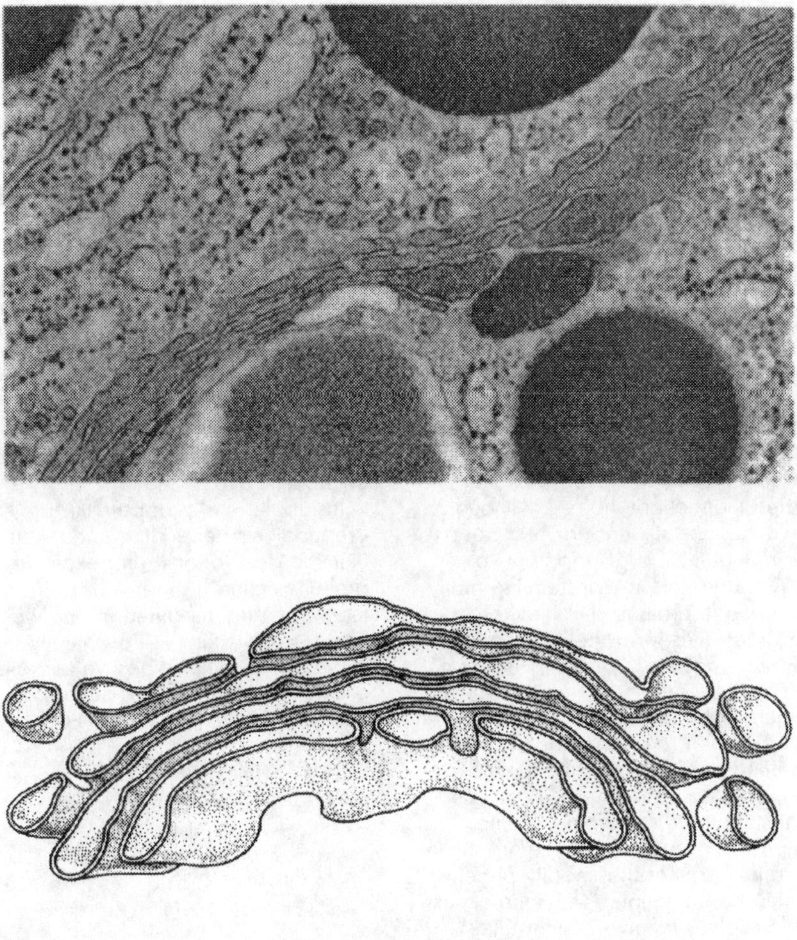

Electron micrograph (top) and drawing (bottom) show a layer of cup-shaped sacs making up a Golgi apparatus.

THE GOLGI APPARATUS, FINAL PROTIEN SORTER

In 1898, the Italian scientist Camillo Golgi, who had been studying stained owl and cat nerve cells under his light microscope, saw a cell structure that did not look like the nucleus. Although some biologists at the time thought the structure was just an artifact—perhaps related to the stains Golgi had used—Golgi believed the newly found organelle played a role in protein secretion.

In the 1960's, Palade and his colleagues confirmed Golgi's theory by using radioactive labeling, staining, and electron microscopy to follow proteins in pancreatic cells as they moved from the rough endoplasmic reticulum, through the Golgi apparatus, and into the secretory granules that carried them out of the cell. Although the details are still unclear, it is now known that the Golgi apparatus plays an important role in transforming many newly made proteins into mature, functional ones by "labeling" them with chemical tags and "shipping" them to their destinations.

Each Golgi apparatus consists of a stack of flat, membranous sacs that are piled one on top of the other like dinner plates. The stack is composed of at least three chemically distinct regions, and each sac in the organelle contains enzymes that modify proteins as they pass through. The sacs closest to the nucleus receive vesicles filled with protein molecules from the endoplasmic reticulum. The proteins must pass through all the sacs in sequence to be processed correctly.

Why does the cell go to such elaborate lengths to modify and sort proteins? According to James Rothman of Princeton University, the compartmental organization of the Golgi apparatus is "crucial to the functioning of the cell, because without it thousands of enzymes would be randomly mixed, resulting in a chaotic splay of biochemical activity." The Golgi apparatus controls the chaos by packaging proteins into vesicles as they pass through the organelle. The protein-filled vesicles then migrate to another organelle or to the cell's surface, where they fuse with the cell's outer membrane and release their contents. This is how cells secrete hormones, enzymes, and other types of proteins as needed.

Inside the Golgi apparatus, proteins are modified depending upon their ultimate destinations. For example, proteins bound for the cell's surface membrane are modified in one way, while proteins that will eventually move to the cell's "digestive" organelles undergo a different series of changes.

LYSOSOMES AND PEROXISOMES, THE CELL'S "DIGESTIVE SYSTEM"

When a white blood cell engulfs a bacterium and destroys it, the white cell's lysosomes do most of the work. They fuse with the vesicle of engulfed material and release digestive enzymes to break up the material. Similarly, when a cell takes in large molecules of food, enzymes in the lysosomes break the food down into smaller and simpler products that the cell can use. These products diffuse through the lysosomes' membranes and go into the rest of the cell, where they serve as building blocks for various structures, until nothing is left inside the lysosomes but indigestible material and the lysosomes become what are called residual bodies. In some cells, the residual bodies then migrate to the cell surface and eject the undigested material into the external environment.

Lysosomes were discovered by a Belgian researcher, Christian de Duve, in 1949, when he homogenized some animal cells and separated them into various components by running them through an ultracentrifuge. After one of these components had been left standing for a few days, de Duve noticed that the level of a certain enzyme in it rose dramatically. Since this enzyme had not attacked any part of the cells before they were ground up, he reasoned that it must have been kept segregated within the cell—probably inside some kind of organelle. He also knew that he had used a relatively gentle method of homogenization, which could have allowed

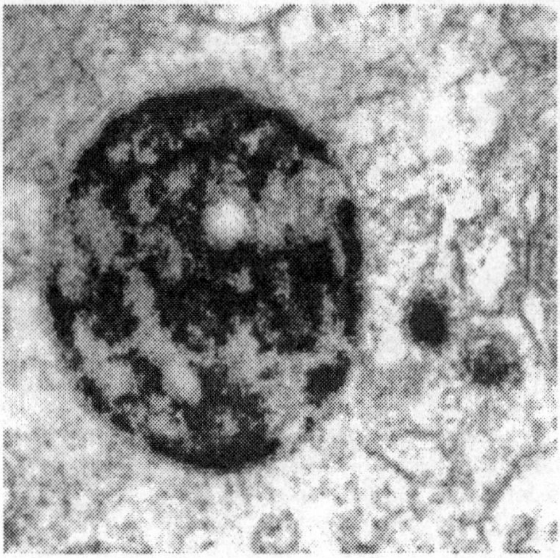

An electron micrograph showing two small lysosomes and one large lysosome. These organelles contain enzymes capable of breaking down various substances.

the unknown organelle to remain intact. Presumably, it released its contents later.

De Duve's biochemical approach, for which he shared the Nobel Prize with Claude and Palade in 1974, was soon supplemented by electron microscopy. But it proved difficult to identify the new particles, since, unlike other organelles, lysosomes vary in shape from cell to cell. Finally, in 1955, Alex Novikoff of the Albert Einstein College of Medicine clearly identified some lysosomes in rat liver cells, and it is now known that lysosomes (whose name refers to the fact that their enzymes can lyse, or digest, substances) exist in all eukaryotic cells.

At about the same time that de Duve and his colleagues were describing the biochemistry of lysosomes, they detected another enzyme-containing organelle. In 1965, de Duve proposed that the organelle be called a peroxisome because it appeared to both generate and break down hydrogen peroxide, a corrosive molecule composed of two atoms each of hydrogen and oxygen.

Today it is known that peroxisomes exist in most eukaryotic cells, and that they are especially prominent in mammalian liver cells. The membrane that surrounds a peroxisome is unusually permeable, permitting many small molecules to enter easily. Peroxisomal enzymes remove hydrogen atoms from these small molecules and join the hydrogen to atoms of oxygen to form hydrogen peroxide. One of the peroxisomal enzymes, catalase, then neutralizes the hydrogen peroxide by breaking it down into water and oxygen. This two-step process is the method that peroxisomes in the liver use to break down molecules of alcohol into substances that can be eliminated from the body. About one-quarter of the alcohol that enters the liver is processed in peroxisomes.

In his early descriptions of peroxisomes, de Duve called them "fossil organelles" because of their primitive nature and seemingly expendable actions. (All of the enzymes found in peroxisomes are also found elsewhere in the cell.) However, it is now known that a rare, fatal genetic disorder called Zellweger's syndrome is the result of malformed peroxisomes, indicating that peroxisomes do have a vital role in the healthy cell.

The ability of peroxisomes to use oxygen in chemical reactions has led many scientists to conjecture that these organelles represent a relic of an attempt by the precursors of eukaryotic cells to "cope" with oxygen as it accumulated in the prehistoric atmosphere. Peroxisomes cannot, however, couple oxygen use with energy production. That ability is restricted to chloroplasts and mitochondria—the "energy converters" of eukaryotic cells.

Lysosomes in Health and Disease

Lysosomes are known to contain over 40 different enyzmes that can digest almost anything in the cell, including proteins, RNA, DNA, and carbohydrates. These enzymes work best in environments more acidic than that found in the cytoplasm, and lysosomes are specially equipped to provide this acidic environment. As corrosive as these enzymes are, they do not ordinarily damage the cell because the lysosomal membrane remains intact.

When cells are programmed to die in some normal process of embryonic development, however—for example, in the metamorphosis of insects—the lysosomes' membranes become permeable and release their enzymes to digest the cells from within. In very old cells, too, the lysosomes may release their contents, which destroy the cell. A more limited form of "autodigestion" can also occur in cells that have been injured by lack of oxygen, an excess of vitamin A, exposure to certain cancer-causing agents, or starvation. In these situations, lysosomes break down a portion of the cell's contents, liberating amino acids that can be used to make the most essential substances and ensuring the cell's survival without major damage.

Lysosomal membranes prevent enzymes from leaving the organelle, but permit entry of new enzymes if they have been properly labeled in the Golgi apparatus. However, if a lysosomal enzyme is not produced or if an enzyme is not properly "addressed" in the Golgi, a lysosomal storage disease can result. Persons with the lysosomal storage disease known as Hurler's syndrome, for example, cannot break down large molecules of sugar-fat compounds called glycosaminoglycans because their lysosomes do not contain the enzyme iduronidase. Glycosaminoglycans accumulate in the lysosomes, swelling them so much that the functioning of the entire cell is impaired.

A particularly severe lysosomal disorder is known as I-cell disease. Children born with this disease lack the entire range of lysosomal enzymes; the enzymes are made, but they are dumped outside the cell instead of being sent to the lysosomes. Various cellular nutrients thus cannot be digested and so pile up in dark lumps, called inclusion bodies, within the lysosomes. The disease affects the kidneys, heart, and nervous system, and children with it usually die of heart failure or pneumonia before reaching puberty.

In the early 1970's, Elizabeth Neufeld, who was then at NIH, showed that the lysosomal enzymes of persons with I-cell disease emerge from the Golgi apparatus without the chemical tag they need to be directed to the lysosomes. She also showed that the defect could be corrected in test-tube cultures of cells taken from people with the disease. The corrective factors she supplied were the specific, properly tagged enzymes that the cells lacked.

Scientists hoped that such "enzyme replacement therapy" could be used to treat people with enzyme disorders. To date, however, it has proved difficult to deliver the missing enzymes to the cells that need them and to induce the lysosomes in each cell to take up the enzymes. When purified enzymes are injected directly into the body, they tend to be quickly destroyed or inactivated. It is particularly difficult to get enzymes into the brain—an important problem now under investigation, since several lysosomal diseases produce severe mental retardation.

In many ways, then, health and disease depend on the lysosomal membrane's ability to control the uptake and release of its contents. Some substances seem to stabilize and strengthen the membranous envelope, while others weaken it. In the future researchers may find ways to use these properties for the prevention of disease or for therapy.

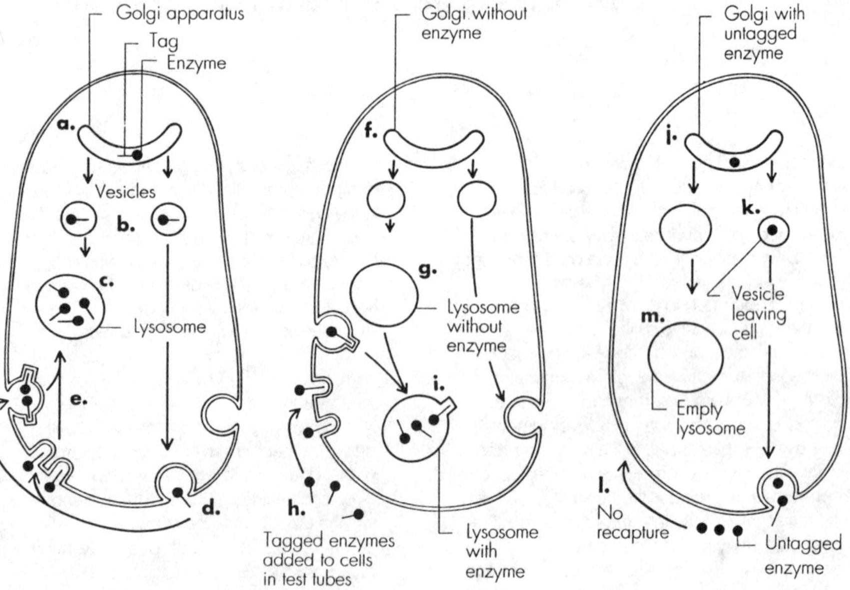

Normal Cell **Hurler's syndrome** **I-cell disease**

The sorting of lysosomal enzymes in a normal cell is shown on the left. First, enzymes receive a chemical "address label" in the Golgi apparatus (a), and move in vesicles (b) to a lysosome (c). Any enzyme that is accidentally swept out of the cell (d) is recaptured and taken back to the lysosome (e). If some part of this process goes awry, a lysosomal storage disorder can result. For example, in Hurler's syndrome (center), an enzyme is not produced (f), and the lysosome therefore lacks that enzyme (g). If a correctly tagged enzyme is added to the cells in test tubes (h), the cells can capture the enzyme and take it to the lysosome (i). In I-cell disease (right), enzymes are correctly made, but they are not tagged in the Golgi (j), and therefore are not sent to the lysosome. When such enzymes leave the cell (k), they cannot be recaptured (l) and so the lysosome remains empty (m).

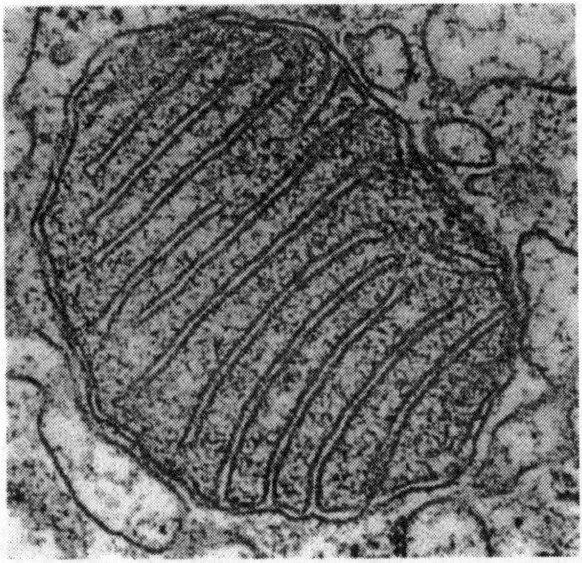

Electron micrograph showing one of the cell's many mitochondria, the organelles that convert energy from food into a form that can be stored.

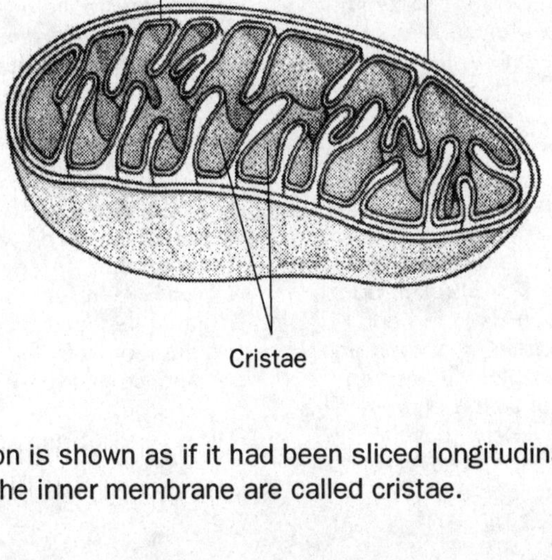

A mitochondrion is shown as if it had been sliced longitudinally. The inward folds of the inner membrane are called cristae.

MITOCHONDRIA, ENERGY CONVERTERS IN THE CELL

One and a half billion years ago, scientists believe, cells derived the energy they needed through a variety of relatively inefficient processes, none of which required oxygen. Oxygen, a waste product of some of these processes, gradually began to accumulate in the atmosphere. It was at this time, scientists hypothesize, that a pre-eukaryotic cell engulfed another primitive cell that had somehow acquired the ability to use oxygen to produce large quantities of energy. Over the eons, a symbiotic relationship evolved between the cells, and today all plant and animal cells have organelles that are the descendants of the primordial energy producers. In animal cells, these organelles are called mitochondria, while the energy-producing organelles in green plants are called chloroplasts.

Chloroplasts use the energy in sunlight to convert molecules of carbon dioxide and water into molecules of sugar, a form of energy that can be stored in the plant cell. (Molecular oxygen is given off as a byproduct of this process, which is called photosynthesis.) When an animal eats a plant (or another animal that has itself eaten plants), the plant's sugars are broken back down into carbon dioxide and water, with the help of oxygen and an arsenal of enzymes, releasing large amounts of stored energy. This energy is immediately converted to yet another form—molecules of adenosine triphosphate (ATP).

ATP is often called the universal currency of cellular energy. It is a convenient way for cells to store the energy they need for such processes as protein manufacture, DNA replication, and the construction of new organelles. ATP is also required for such mechanical work as muscle contraction, pumping water through membranes, and cell movement. Except for the first stage of sugar breakdown, the entire, complicated process of energy transfer from sugar to ATP takes place within the animal cell's mitochondria. (Plants also have mitochondria, although they make most of the ATP they need within their chloroplasts.)

Besides supplying energy, mitochondria apparently help to control the concentration of water, calcium, and other charged particles (ions) in the cytoplasm. They also break down and recycle the energy contained in fatty acids and amino acids.

Mitochondria are the largest organelles in an animal cell, after the nucleus, yet some cells have more than a thousand of them. They vary in diameter from 0.5 to 1 micrometer and in length up to 7 micrometers, and can be seen with a good light microscope. Mitochondria are usually represented as oval shaped, but in life they can change shape quite readily. They swell or contract in response to various hormones and drugs and during ATP manufacture. This swelling and contracting appears related to the movement of water through cells, and is particularly evident in the kidneys, through which 180 liters of water are filtered daily.

Although mitochondria were first observed in the 1880's, it took many years for scientists to understand the organelles' function. The process by which mitochondria use oxygen to release the chemical energy stored in food is called cellular respiration. Early in this century, it was discovered that the biochemical reactions of respiration fall into two main groups: the carbon pathway, in which sugar is broken down into carbon dioxide and hydrogen; and the hydrogen pathway, which transfers hydrogen to oxygen in stages, forming water and releasing energy.

In the hydrogen pathway, the hydrogen's electrons pass through an "electron transport chain" made up of enzymes. As they move from enzyme to enzyme, the electrons give up part of their energy. This energy is then stored in molecules of ATP. In the end, 38 molecules of ATP are formed for every molecule of sugar that is used up in respiration.

Mitochondria are marvelously efficient at converting the chemical energy of sugar into ATP. Whereas a man-made engine would be considered very efficient if it converted 25 percent of the energy available in gasoline into mechanical work, mitochondria routinely turn 54 percent of the available energy in sugar into ATP. This efficiency is achieved, in large part, because of the mitochondria's internal structure. In the early 1950's, Palade and a Swedish scientist, Fritiöf Sjostrand, reported that mitochondria are bounded by a membrane and that they have a system of parallel, regularly spaced inner ridges that the scientists named cristae. It is now known that there are two membranes around a mitochondrion: an outer membrane, separated from the rest of the organelle by a fluid-filled gap; and an inner membrane that is folded inward at various points to increase its surface, forming the cristae. This ridged surface allows the enzymes of the electron transport chain, which are attached to the cristae, to be packed more densely within each mitochondrion, thus increasing the organelle's efficiency. This general design seems to have existed unchanged from the time that mitochondria-like cells were free-living organisms.

Mitochondria have also kept other vestiges of their existence as independent organisms. For example, mitochondria "reproduce" by splitting in half as many modern bacteria do; they are not formed by budding from existing cellular structures or built up from simple cellular constituents, as is the case for ribosomes.

More significantly, after a billion or so years of residence within "host" cells, mitochondria (and chloroplasts) still retain some of their own DNA. The amount of this non-nuclear DNA varies significantly from organism to organism. The chloroplasts of plants, for example, have five times more DNA than do the mitochondria of mammalian cells. Human mitochondrial DNA is a closed, circular molecule 16,569 nucleotide pairs long. Although this is less than 1 percent of the total DNA in a human cell, each mitochondrion has enough to code for several of the key inner membrane proteins. (All of the other proteins in a mitochondrion are coded for in the nucleus, made on free ribosomes in the cytoplasm, and imported into the organelle.)

Another curious characteristic of human mitochondria is the fact that all of a person's mitochondria are descendants of those of his or her mother; no paternal mitochondria are present. This fact has proved useful to evolutionary biologists, who can study the passage of mitochondrial DNA from generation to generation while ignoring the "interfering" information contained in the nuclear DNA, which records the genetic contributions of both parents.

Scientists have long suspected that defects in mitochondrial genes could lead to inherited disease in the same way that mistakes in nuclear DNA do. This hunch was not proven until 1988, when Douglas Wallace of Emory University showed that a rare eye disease called Leber's hereditary optic neuropathy is caused by a mutation in mitochondrial DNA. The defective mitochondrial gene prevents the optic nerves from producing enough ATP, and the nerves, which need huge amounts of ATP and are thus particularly sensitive to any deprivation, die. When he announced these findings, Wallace said, "we feel that these alterations [in mitochondrial DNA] may be responsible for a wide spectrum of diseases in the brain, the central nervous system, and the musculoskeletal system."

Mitochondria, chloroplasts, and the other organelles described thus far are surrounded by membranes. But cells can also contain threadlike organelles that lack membranes. These extremely fine structures serve as buttresses, highways, and motile mechanisms for the cell.

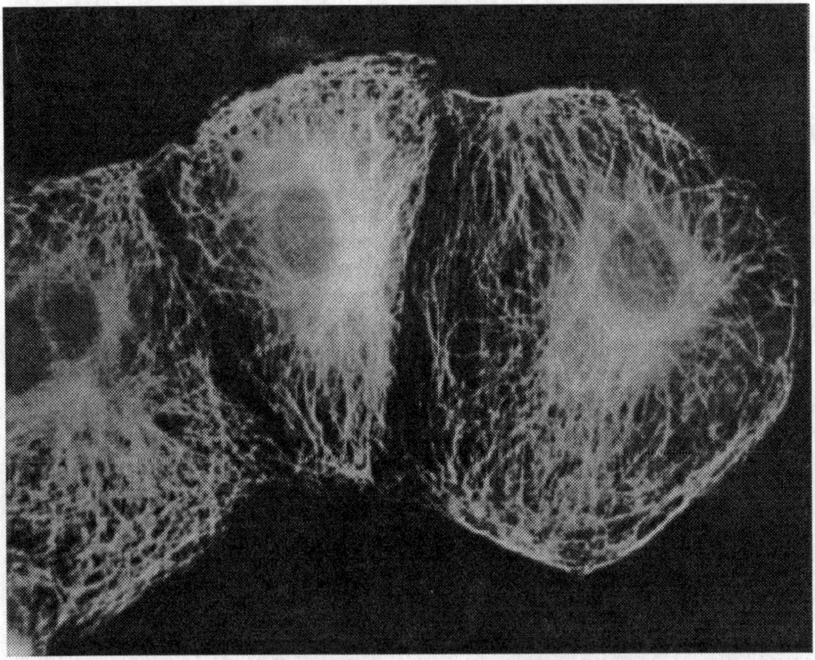

These cells have been stained with modified antibodies that attach to the cell's "scaffolding," called the cytoskeleton, and that glow under ultraviolet light.

THE CYTOSKELETON, THE CELL'S PHYSICAL PROPS

Many cells in a multicellular organism must combine the seemingly contradictory traits of stability and mobility. With few exceptions, multicellular organisms begin to develop when a motile sperm meets an egg. Many cell divisions occur, and then cells migrate to their final positions. During life, individual cells divide frequently, and certain specialized cells move through the body to accomplish various tasks. In addition, every cell must have a mechanism for moving materials within itself. Balancing the need for movement is the requirement for cell stability. A cell must maintain its shape against the pressure of surrounding cells. Keeping a cell firm while enabling it to move are the twin roles played by the cytoskeleton.

For a long time, microscopists believed that the cytoplasm surrounding the cell's organelles was completely unstructured. But as scientists began to use newer and gentler fixatives to prepare cells for electron microscopy, a lacy network of fibers was revealed. These structures crisscross the cell like girders and it was hypothesized (and later shown experimentally) that, like an animal's bony skeleton, these structures play a role in giving the cell its shape. For this reason, they are known collectively as the cytoskeleton.

There are three main kinds of cytoskeletal fibers—microfilaments, microtubules, and intermediate filaments—which are distinguishable both by their structure and by their protein composition. All three support and stiffen the cell. In addition to their structural roles, microtubules and microfilaments are essential for a variety of dynamic whole-cell activities, including division, contraction, and crawling, as well as for the movement of vesicles within the cell.

Microfilaments are more commonly called actin filaments because they are composed of "beads" of the protein actin arranged into long, slender chains. Each filament is only 6 nanometers in diameter; they are the finest of the cytoskeletal components. The role that actin filaments play in muscle contraction has been thoroughly studied over the past 30 years. In the 1950's, a British scientist, Hugh Huxley, proposed a model for muscle contraction that has since been shown to be correct. According to the model, each muscle cell comprises parallel rows of actin filaments that alternate with rows of another protein, myosin. When stimulated by an influx of calcium, projecting "arms" of myosin "grab" the adjacent actin filaments and pull, causing the muscle cell to shorten. Contraction is an ATP-requiring process; each "grab" and release uses up one molecule of ATP. In recent years, researchers have found evidence of similar actin-myosin interactions in many other kinds of cells, including cells that secrete hormones and white blood cells that move through the body to fight invading organisms.

Microtubules, at 22 nanometers in diameter, are the thickest of the cytoskeletal components. They were

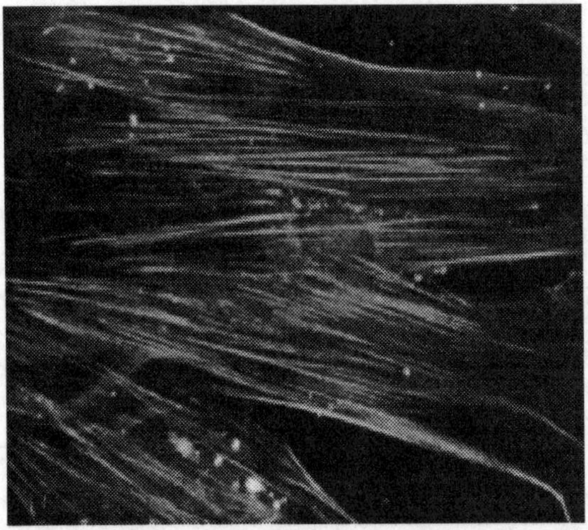

The microfilament bundles in this skin cell have been stained with modified antibodies that glow under ultraviolet light.

noticed in the mid-1950's, but were seen only rarely until 1963, when the gentle fixative glutaraldehyde was developed. Each hollow tubule is composed chiefly of small, spherical subunits of proteins called tubulins. Microtubules assemble spontaneously from "pools" of tubulin when needed and, under appropriate conditions, dissolve, or depolymerize, back into their tubulin subunits. (Microfilaments also form and break down spontaneously.) Under the microscope, microtubules can be observed growing and shrinking rapidly. Because microtubules perform so many important functions within the cell, scientists are eager to learn the details of their dynamics.

One of the most vital functions of microtubules is to aid in cell division. Just before a cell divides, small bodies called centrioles (which are themselves composed of microtubules) migrate to the cell's poles. A spindle made of microtubules forms between the centrioles. Chromosomes attach to the spindle, which then helps to guide them to the daughter cells. In 1988, Marc Kirschner and his colleagues at the University of California, San Francisco, found strong evidence that chromosomes move toward the poles as the microtubules slowly dissolve. This indicates that, contrary to what many researchers had assumed, microtubule depolymerization and subsequent chromosome movement apparently do not require ATP.

The spindle is easily disrupted by a number of chemicals. Colchicine, an extract of meadow saffron used

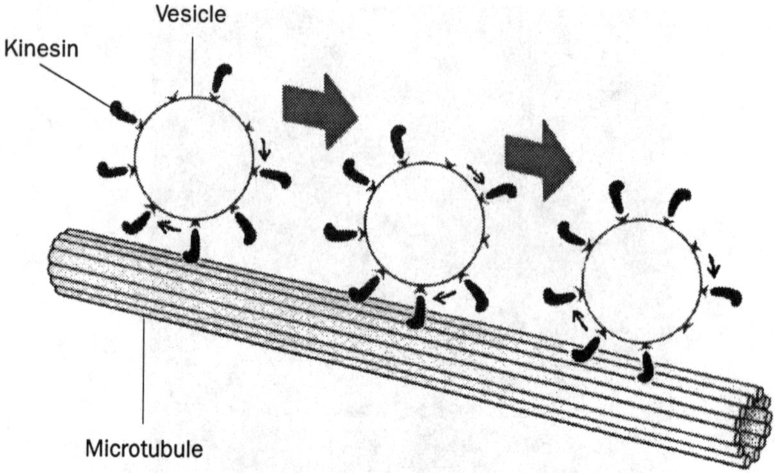

Molecules of kinesin connect a vesicle to a microtubule (left) and flex (center), thus pulling the vesicle along the microtubule (right).

since ancient Egyptian times as a treatment for gout, stops spindle formation and arrests cell division. In 1967, Edwin Taylor and his associates at the University of Chicago discovered that colchicine did its work by binding to tubulin. Two other chemicals, vinblastine and vincristine, also disrupt spindles and are used as anticancer drugs since they preferentially inhibit the spindles of rapidly dividing cancer cells.

Between cell divisions, microtubules act as miniature highways along which vesicles carrying such materials as hormones, neurotransmitters, and nutrients move. Using new techniques, such as video-enhanced light microscopy, scientists in several laboratories have observed microtubules interacting with a globular protein called kinesin that functions as a molecular motor to move vesicles and organelles along microtubule tracks toward the cell surface. Kinesin also moves vesicles filled with neurotransmitters along the microtubules within nerve cell axons. A motor protein that moves vesicles in the opposite direction, toward the cell's interior, was discovered in 1987.

Microtubules are also involved in the movement of cilia and flagella. These whiplike filaments project from certain cells and perform a variety of tasks. Large numbers of cilia are found on cells that line the respiratory system, for instance, where they help to sweep out dust and debris. Both cilia and flagella play an important role in human reproduction. The coordinated beating of cilia in the

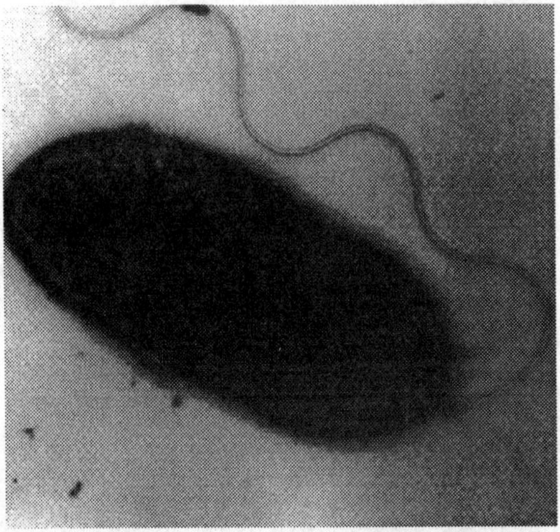

Some microorganisms are equipped with a flagellum (composed of microtubules), which thrashes to propel the animal.

oviduct produces a sort of current that draws the egg into the uterus, while the rapidly thrashing flagella of sperm help them to "swim" toward the egg.

The inherent ability of microtubules and microfilaments to assemble and disassemble rapidly allows for the construction and destruction of these cytoskeletal components to suit the needs of a moving cell. In contrast, intermediate filaments are the most stable of the cytoskeletal fibers. At 8 to 10 nanometers in diameter, they are intermediate in size between microfilaments and microtubules. Intermediate filaments appear in many types of cells, and their precise protein composition depends upon their location. They are prominent in parts of the cell that are under mechanical stress, such as the long axons of nerve cells and the surface of skin cells.

Some researchers have reported that a number of diseases, including Alzheimer's disease and Parkinson's disease, are associated with changes in the intermediate filament arrays in cells. However, while intermediate filaments obviously have important functions in cell physiology, very little is now known about the cause or effect of their alteration during disease.

Nevertheless, an understanding of intermediate filaments is proving valuable in tumor diagnosis. If cancerous cells arise in one organ and then migrate to another, a clinician can study the intermediate filament composition of the tumor cells to determine the cancer's origin. This knowledge helps physicians decide on the best course of treatment for that particular tumor.

Some scientists believe that intermediate filaments may play a role in carrying certain messages from the cell's surface to the nucleus, where they turn genes on. Research is under way to discover if this is, in fact, the case.

Current investigations of the cell's organelles—the nucleus, ribosomes, endoplasmic reticulum, Golgi, lysosomes, peroxisomes, mitochondria, and cytoskeleton—hold great promise for the solution of problems of basic biology and clinical medicine. However, the key that may unlock the greatest number of health benefits may well be found in the cell's filmy membranes, particularly the surface membrane, which plays a pivotal role in maintaining the integrity of the cell and, in a larger way, in protecting the health of the organism.

THE SURFACE MEMBRANE, VERSATILE GATEKEEPER

"In the beginning," writes Gerald Weissmann of New York University, "there must have been a membrane! Whatever flash of lightning there was that organized purines, pyrimidines, and amino acids into macromolecules capable of reproducing themselves, it would not have yielded cells but for the organizational trick afforded by the design of a membrane wrapping." Weissmann imagines these primitive membranes forming bubbles in which the first macromolecules were enclosed and protected from dissipation in the salty primordial seas.

A cell's outer membrane is often thought of as a boundary that defines the living cell from its surroundings. And, indeed, surface membranes are crucial in keeping cells intact. Moreover, the internal membranes that wrap around many organelles in eukaryotic cells separate the cytoplasm into discrete regions, somewhat like the walls that form rooms in a house. These inner membranes enable the cell to perform many biochemical activities simultaneously, thereby greatly increasing the cell's efficiency.

Yet despite its barrier functions, the cell membrane—which is often less than 0.01 micrometer thick—is not impassive. Rather, it is exquisitely sensitive to its surroundings and selectively allows certain substances to enter and leave the cell while barring others. It takes in nutrients and excretes wastes. It sends and receives chemical and electrical messages, including signals for the cell to manufacture proteins or to divide. In multicellular organisms, it joins with other cells to form tissues.

These myriad abilities are due to the membrane's composition. Although surface membranes differ in their precise composition depending on the cell's type, and although a membrane's configuration changes from moment to moment, all membranes are composed of two basic kinds of molecules—proteins and lipids (fats).

In 1972, S. Jonathan Singer and Garth Nicolson of the University of California, San Diego, proposed a model to describe the relationship of proteins and lipids in an idealized membrane. They compared the proteins to "icebergs floating in a sea of lipids," and suggested that some of the proteins are folded so that the "tips" poke above and below the plane of the membrane, while the middle of the protein is embedded in the membrane itself.

Although such tripartite proteins were unknown at the time Singer and Nicolson proposed their model, they have since been shown to exist. Many other membrane proteins that are attached either to the inner or outer face of the surface membrane have also been studied in detail in the years since Singer and Nicolson proposed their so-called fluid-mosaic model. It is now known that the proteins do not float through the lipids unrestrained. Rather, they are tethered in a general region by slender

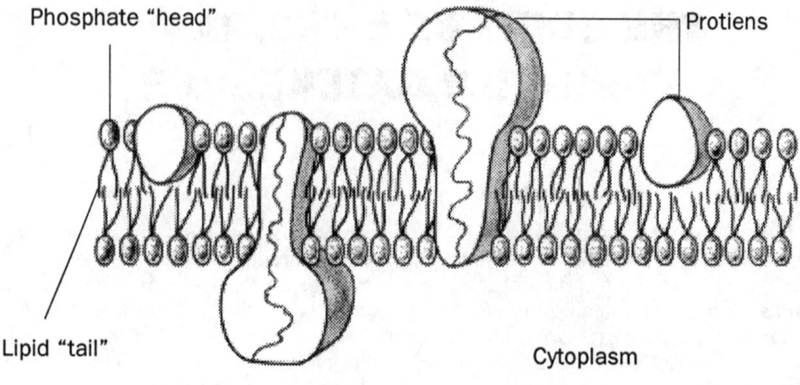

The surface membrane of a cell consists of a double layer of lipid molecules interspersed with protein molecules. Surface membranes play many roles, including keeping the cell intact and allowing appropriate substances to enter and leave the cell.

fibers somewhat like hot-air balloons are tethered by guy wires before liftoff.

The lipids that make up the bulk of a cell's surface membrane fall into three classes: phospholipids, steroids (primarily cholesterol), and glycolipids. About half of the molecules in an average membrane are phospholipids. Each phospholipid molecule has a water-seeking (or hydrophilic) phosphate "head" and two flexible, water-avoiding (or hydrophobic) lipid "tails." In a surface membrane, phospholipids spontaneously arrange themselves into a bilayer (double layer) with phosphate heads touching the watery interior and exterior of the cell, and lipid tails buried in the middle of the layer.

Cholesterol is abundant in many animal cell membranes; sometimes there is as much as one cholesterol molecule for every phospholipid. Cholesterol is a rigid molecule that gives the surface membrane strength. It is manufactured within the cell (on the endoplasmic reticulum) and is also brought into the cell from the blood. Cholesterol is present only in animal cells; plant cells are stiffened by a very rigid cell wall composed mainly of cellulose.

Glycolipids are composed of a sugar ("glyco" is derived from the Greek word for sweet) and a lipid portion, and make up about 5 percent of the lipid population. A person's blood group (O, A, B, or AB) is determined by the particular kind of glycolipids present on the surface of his or her red blood cells.

Liposomes – Drug Delivery Vehicles of the Future

Can microscopic artificial membranes help doctors treat cancer, angina, and viral infections more effectively, and lead to better vaccines, bronchodilators, eye drops, and sunscreens? The researchers who are developing liposome technology hope so. A liposome is a tiny sphere of fatty molecules surrounding a watery interior. Because they are made of the same material as cell surface membranes, liposomes stick to cells and are not toxic. These characteristics make them attractive candidates for drug delivery vehicles.

In 1980, two groups of researchers used liposomes filled with a common antibiotic to cure mice having a severe, but localized, infection. The infected cells were of a kind that is specialized to take up foreign bodies, and so they readily engulfed the liposomes. However, getting other kinds of cells to take up drug-filled liposomes has proven to be more difficult. A number of groups of researchers are experimenting with antibody-tagged liposomes filled with an anticancer drug. The liposomes are guided to the diseased tissue by the antibodies, which seek out cancerous cells but spare healthy ones. This selectivity allows smaller amounts of a drug to be used with greater effect, an important advantage considering the serious toxicity of many anticancer drugs.

Other research teams are developing liposome-drug compounds that would be injected into muscle to release growth hormone or anticancer agents over a period of weeks. Scientists also hope to use liposomes to improve the safety and effectiveness of vaccines, including an influenza vaccine. As the cost of both natural lipids (extracted from egg yolk and soybeans) and artificial lipids declines, the future may bring many other liposome-containing medical products as well as nonmedical items, such as cosmetics.

DIRECTING TRAFFIC ACROSS THE SURFACE MEMBRANE

The oily lipids of a cell's surface membrane serve admirably to prevent the cell's water-based contents from leaking out. However, in "solving" this problem, the cell is confronted with another—how to transport wastes and cell products out of the cell and allow nutrients and other substances in, without either shrinking or swelling too much.

Over eons, cells have evolved a wide variety of transport mechanisms to ferry substances across the hydrophobic barrier. Transport may be either "passive," which requires no energy, or "active," which uses ATP Also, a molecule may either pass directly through the lipid layer or it may be carried in by a surface protein in a process called receptor-mediated endocytosis. (Endocytosis can also occur without the involvement of a surface protein. In both styles of endocytosis, a bit of the surface membrane folds inward around the entering particle, then pinches off and carries the particle into the cell. The opposite process, exocytosis, occurs when vesicles moving from the cell's interior fuse with the surface membrane and spill their contents outside of the cell.) The method used to import or export substances depends on a combination of the transported item's size, chemical composition, electrical charge, and abundance (concentration), as well as on its ability to dissolve in lipids.

Oxygen, nitrogen, and other small molecules that can dissolve easily in lipids move readily back and forth across the bilayer. Importantly, because of its small size and the distribution of its electrical charge, a water molecule can also pass relatively easily through the membrane even though water is quite insoluble in oil.

In contrast, large molecules, such as proteins and sugars, cannot pass through the lipid bilayer. A variety of transport systems, many of which involve surface proteins, are used to ferry these substances into and out of the cell. Surface membrane lipids are also highly impermeable to all ions, no matter how small they are. ATP-requiring protein "pumps" are employed to transport ions.

One well-studied pump system is the sodium-potassium pump. This membrane protein consumes more than a third of the cell's total ATP production in an endless cycle of pumping sodium ions out of the cell while drawing potassium ions in. This frantic work results in an "ion gradient" in which there is a high concentration of sodium ions outside the cell and a high concentration of potassium ions inside. This creates a source of potential energy analogous to the potential energy created when water is held behind a dam. If the dam is lowered, water will flow over it and can be used to turn turbines or do other work. In the same way, if the membrane pumps momentarily permit an inward flux of sodium ions (accompanied by an outward rush of potassium ions), a variety of tasks,

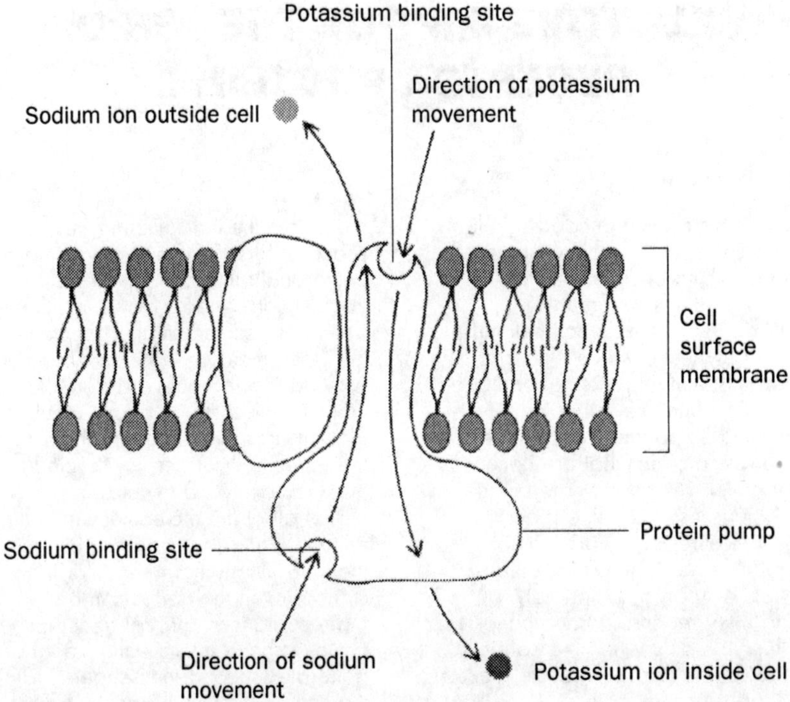

This schematic drawing shows how the sodium-potassium ion protein pump in the cell's membrane forces sodium out of the cell, while pulling in potassium.

including the propagation of electrical signals among nerve cells, can be performed.

Another very important function of the sodium-potassium pump is to keep the total number of ions on both sides of the membrane approximately equal, thereby preventing the cell from swelling or shrinking. Ouabain, a crystalline substance derived from the seeds of an African shrub, disrupts the sodium-potassium pump and causes cells to swell and burst. African hunters use it in large amounts to make poisoned darts, and doctors sometimes prescribe it in small amounts as a heart stimulant.

SELECTIVE IMPORT, THE JOB OF RECEPTOR PROTEINS

Animal cells are surrounded by blood that contains many vital substances, some of which are present only in minute quantities. From this rich blend of ingredients, each cell must extract only those things it needs, and reject anything it is not equipped to handle. Such selective import is performed by specialized surface membrane proteins that are collectively called receptors. The unique characteristics of a cell depend, in large measure, on what kinds of receptors it has. Like a lock that accepts only an appropriately shaped key, each different receptor will function only when the correctly shaped blood-borne molecule (called a ligand) attaches to it.

One important receptor is the LDL (low-density lipoprotein) receptor, which admits cholesterol into the cell. When a normal cell needs more cholesterol than it has produced, it synthesizes more LDL receptors. The liver manufactures cholesterol and releases it into the blood, and it is carried by LDL's to the waiting LDL receptors. When the LDL's enter the cell, lysosomes break them down and release the cholesterol. As soon as the cholesterol in the cytoplasm reaches a certain concentration, the cell stops making new LDL receptors.

LDL receptors bound to LDL's do not enter the cell one at a time. Rather, the complexes are rounded up and migrate to a place on the membrane where the surface begins to sink inward. These places, called coated pits, eventually pinch off and move into the cell. During this process, bits of the surface membrane are brought into the cell along with the LDL complexes. However, the cell does not become smaller, because the "lost" membrane is replaced by vesicles moving outward from the cell's interior. It has been calculated that an area of membrane the size of the entire cell surface is recycled in this way every 50 minutes.

Not all of the molecules that influence cell function are absorbed by the cell. Many hormones, for instance, exert their effects in an indirect manner, through receptor proteins that transduce the signal and generate "second messengers" within the cell. One of the best understood of these second messenger systems employs proteins called G proteins because they add phosphate atoms to a substance that contains guanine.

In the G protein system, when a "first messenger" (such as a hormone) reaches the cell surface, it binds to a receptor that then sends a signal to a G protein located on the cytoplasmic side of the cell membrane. Depending on its type, the activated G protein then either stimulates or inhibits the activity of any of a number of enzymes, including one called adenylate cyclase. This enzyme causes cyclic AMP, a common second messenger, to be produced. Cyclic AMP then sets off a chain reaction that eventually results in changes in the shapes of certain proteins in the cell, which, in turn, lead to still other cellular responses. One of these

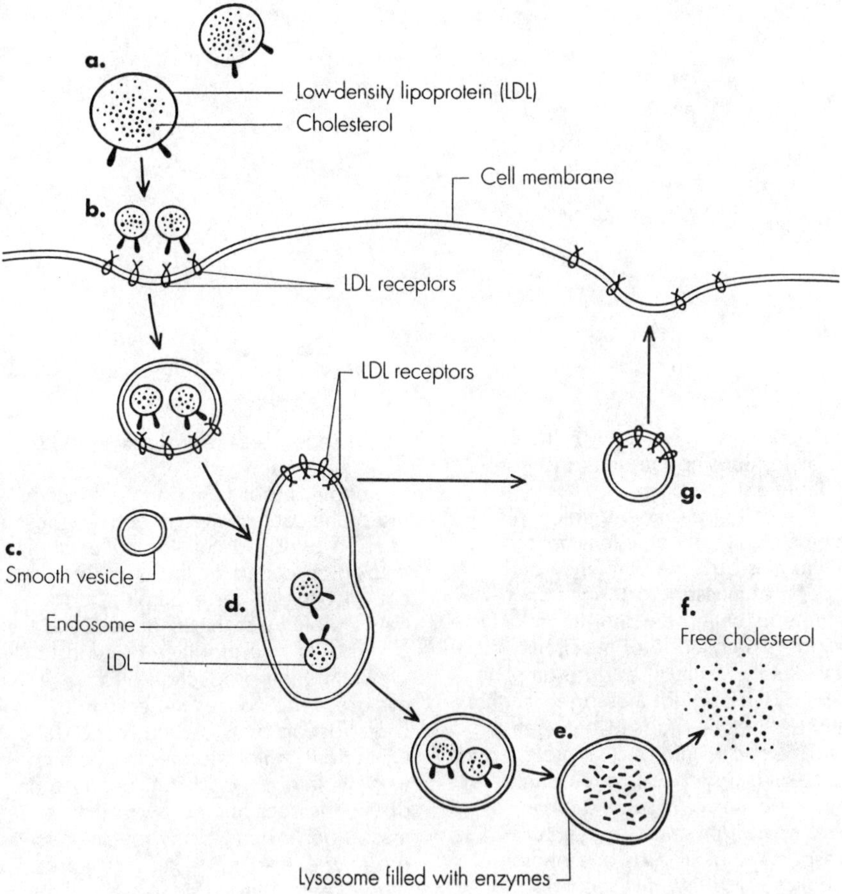

Steps in the endocytosis of cholesterol. Low-density lipoproteins (LDLs) filled with cholesterol approach the cell (a) and attach to LDL receptors in the cell's surface membrane (b). After passing into the cell, a vesicle filled with LDLs joins with a smooth vesicle (c) to become an endosome (d), in which the LDLs and the LDL receptors pull apart. A vesicle filled with LDLs migrates to a lysosome (e), which breaks down the LDLs and releases free cholesterol into the cytoplasm (f). Meanwhile, the LDL receptors are returned to the cell surface (g).

responses is to "switch off" the G protein, causing the message transmittal to stop.

The cell appears to employ this complex signaling system because it increases both the efficiency and speed of message transmission. A single incoming messenger molecule triggers a cascade of reactions that eventually results in a large amplification of the original message. Furthermore, the time elapsed between the arrival of a signal at a G protein and a cellular response is often only a few fractions of a second. For example, light-sensitive eye cells respond to as little as one photon of light in just a few milliseconds through a second messenger system. In contrast, other cells take as long as 30 seconds to respond to signals from the environment.

Certain diseases impair the functioning of the second messenger system, causing profound cellular malfunction. A toxin produced by the organism that causes cholera, for example, "locks" the G proteins of intestinal cells into the "on" position so that they are constantly stimulating the production of adenylate cyclase. This causes the cells to excrete vast amounts of fluid, accounting for the often-fatal diarrhea associated with cholera. In addition, because cell division is coordinated by a second messenger system, many researchers are now attempting to discern how a breakdown in this system might allow a cell to become cancerous.

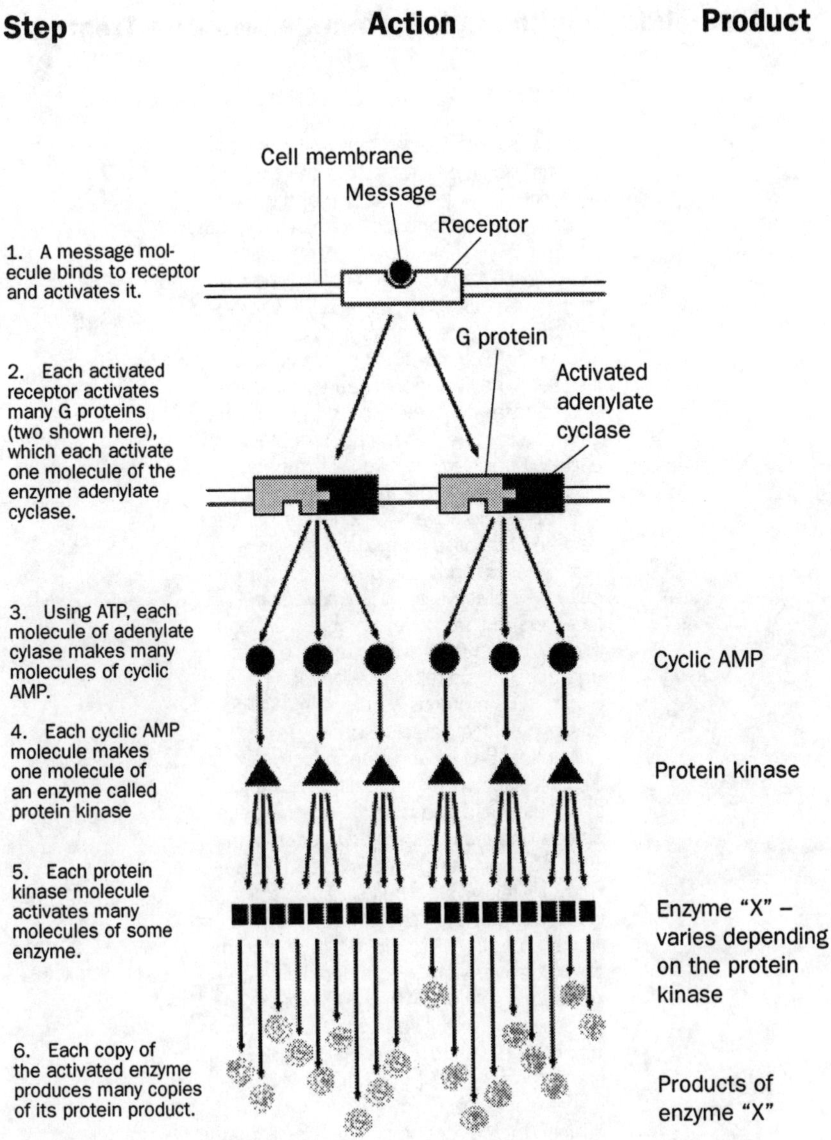

How a signal brought to the cell by a single message molecule is amplified through the second messenger system.

A Breakdown in the LDL System Can Now Be Treated

The low-density lipoprotein system works like a thermostat to ensure that the cell always has enough cholesterol for life and growth, without accumulating too much of it. Because of a defective gene, however, the cells of people with familial hypercholesterolemia (FH) are severely deficient in LDL receptors. This breaks the cells' normal chain of control and leads to over-production of cholesterol by the liver. The liver manufactures cholesterol and releases it into the blood, but since the cells do not have enough LDL receptors, not enough cholesterol is taken up. Instead it begins to clog the arteries.

People who have two genes for FH suffer from heart disease from a very early age, sometimes as early as age 1. Persons with only one defective gene do make some LDL receptors, but still have a high risk of heart attacks, and often have an attack before age 30. FH is a relatively common genetic disorder; as many as 1 person in 500 carries one copy of the defective gene.

Michael Brown and Joseph Goldstein of the University of Texas Health Science Center at Dallas elucidated both the precise genetic defects and the role of the LDL receptor in FH. In 1985, both men were awarded Nobel Prizes for this work. Their studies also led to the development of lovastatin, a drug to treat FH. Lovastatin has a dual action: it damps down the liver's ability to manufacture cholesterol so that fewer LDL's enter the bloodstream, and it causes the LDL receptors that the person has to take up cholesterol more efficiently.

The mechanism Brown and Goldstein outlined for LDL receptors is but one example of the process of receptor-mediated endocytosis. The cell also uses protein receptors to take in such substances as insulin, transferrin (an iron-bearing compound), and cell complexes that are produced when the immune system is activated.

A GLIMPSE OF THE FUTURE

In the three centuries since Robert Hooke turned his microscope on bits of dried cork, much has been learned about the world inside the cell. Directed by the genes and influenced by the environment, cells perform an astonishing array of tasks and take on a variety of forms suited to their work. Cell biologists now know a great deal about how the cell's living machinery works to make proteins, how many of the organelles are formed, and the steps involved in cell division. With so much revealed, it may seem that there is little mystery left in cells. But this is far from true.

Large portions of the cell's membranes, for example, remain uncharted territory. Although strides have been made in discerning the structure, function, and location of many surface and inner membrane proteins, studies of membrane lipids have lagged behind. In the years ahead, scientists will be attempting to discover how lipids in animal cells (which are manufactured at a single site, the endoplasmic reticulum) are transported to multiple sites throughout the cell.

Studies of cell organelles have been intense in recent decades, but much still remains to be learned about, for example, the actions of the Golgi apparatus and the endoplasmic reticulum. Protein receptors and trafficking signals that direct newly made proteins through these organelles have been postulated, but scientists have yet to isolate and characterize these systems. New techniques will probably be devised for examining the roles of cellular components, but there is still a place for such time-honored methods as cell fractionation, which was used by Paul Lazarow of The Rockefeller University and other investigators to separate peroxisomes from other cell components, thus fueling renewed interest in the study of this organelle.

In the future, isolated studies of cell organelles and molecules will be increasingly supplemented by investigations that aim to probe more global cellular functions. According to Norton Gilula of the Research Institute of Scripps Clinic, "the greatest single challenge in cell biology today is learning what regulates cell division and differentiation." How, for example, does each cell in a developing organism "know" when to turn the appropriate genes "on" or "off"? Why do some cells become "specialists," while others remain "generalists"? Why do normal cells eventually wear out and die—and what factors hasten or slow cell death?

To answer such questions, says Gilula, "we need to know not only details of shape and action for individual molecules, but also how these molecules interact in complete cells." Because the fundamental unit of an organism is the cell, whatever is learned about individual genes, proteins, and other molecules must be reinterpreted at the cellular level before a full understanding of the role of each in health and disease is possible.

Already, basic research on how cells

function has led to a variety of improvements in human health care. For example, an understanding of hormone release, binding, and action gained through the study of cellular models enabled pharmacologists to design a drug for high blood pressure. The drug works by selectively inhibiting the enzyme that normally allows blood pressure to rise.

Another area of basic cellular research that may lead to new therapeutic products is the study of angiogenesis, or blood vessel formation. Since tumors must be well vascularized in order to thrive, agents that inhibit angiogenesis might be employed to combat cancer. Conversely, agents that promote blood vessel growth might be useful as treatments for wounds and burns.

Finally, techniques that were originally developed to study genes and other cell components and to answer basic questions about cell activity are now being applied to medical problems, and offer great promise for the treatment of many illnesses. Technologies for fusing different cell types, for example, led to the development of "hybridomas"—antibody-producing cells fused with tumor cells. Hybridomas are living factories that produce large quantities of a single, specific antibody. These antibodies can detect, with great speed and accuracy, agents that cause such diseases as gonorrhea, hepatitis B, and AIDS.

"Cell studies," says Gilula, "form a bridge between what is known at the genetic level and what needs to be learned before new therapies can be developed." We are indebted to the scientists of the past who first revealed the marvels of the cell. As exciting as the past has been, however, the future promises to be still more thrilling as researchers begin to gain an even deeper understanding of cell activities and to apply that understanding to questions of health and disease.

GLOSSARY

Abzyme An antibody produced in the laboratory that has the catalytic properties of an enzyme.

Active Site The region of an enzyme where substances upon which the enzyme acts bind to it, allowing a chemical reaction to take place.

Alpha Helix A common structural element of proteins. The alpha helix has a backbone consisting of a tightly coiled, right-handed helix.

Amino Acid A building block of proteins. There are 20 different kinds of amino acids; a protein consists of a specific sequence of amino acids.

Angstrom A unit of length, one hundred-millionth of a centimeter (approximately .000000004 inches); commonly used for describing atomic dimensions.

Antibody A protein of the immune system produced by animals in response to an antigen (a foreign, often disease-causing, particle). Antibodies bind uniquely to antigens.

Antigen A substance that stimulates the production of antibodies.

Area Detector An instrument that can distinguish individual X-ray beams scattered by a crystal, allowing scientists to record the intensity from all beams at once. As a result, X-ray diffraction patters that took months to collect can now be obtained in hours, and with greater accuracy.

Atom The smallest particle that can make up an element. Each atom consists of a specific number of protons, neutrons, and electrons.

Atomic Resolution The ability to see a molecule in such fine detail that the arrangement of individual atoms is resolved.

ATP (adenosine triphosphate) The compound that serves as a source of energy for the physiological reactions in cells.

Bacterium A one-celled microorganism that contains no nucleus.

Base The basic subunit of DNA or RNA. Paired bases – adenine with thymine and guanine with cytosine (uracil replaces thymine in RNA) – make up each "rung" of the "ladder" of the DNA molecule. See *nucleotide.*

Beta-Pleated Sheet A common structural element of proteins. The beta-pleated sheet consists of strands of polypeptide chain lying side by side, connected by hydrogen bonds.

Binding Pocket A region of the three-dimensional structure of a protein into which other molecules fit in order to bind.

Biochemistry The study of the chemical reactions that occur in living organisms.

Bond The attraction that causes atoms to "stick together" to form a molecule.

Cell The basic subunit of any living organism; the simplest unit that can exist as an independent living system.

Cell Division The doubling in mass and splitting of one "mother" cell into tow "daughter" cells.

Cell Surface Membrane A complex film of lipids interspersed with proteins that covers the cell. The membrane maintains the integrity of the cell and controls what goes in and what comes out.

Centrifuge A machine that separates particles according to their size and density by spinning them at varying speeds.

Chemical Bond See *bond.*

Chloroplast The chlorophyll-containing organelle in green plants in which light energy is converted into sugars.

Cholesterol A waxy lipid produced by animal cells that is a prominent component of cell membranes.

Chromosome A rod-shaped structure containing genes that is found in the cell nucleus. It is composed of DNA and proteins, and can be seen in a light microscope during some stages of cell division.

Cocrystal A crystal consisting of two bound molecules, such as protein and DNA.

Codon A triplet of nucleotides in the DNA or RNA molecule that codes for 1 of the 20 amino acids in proteins or for a signal to start or stop protein production. Each gene that codes for a protein is a group of codons that give the instructions for building that protein.

Computer Graphics Images generated and manipulated by computers, which are very useful in studying the possible three-dimensional configuration of complex structures like protein molecules.

Cristae The inward folds of a mitochondrion's inner membrane.

Crystal A solid that has a characteristic structure in which individual molecules assume regular positions in relation to one another. While some substances readily form crystals, it is not always easy to "grow" (induce the formation of) crystals of many substances that researchers would like to study by X-ray crystallography.

Cytoplasm All the substance inside a cell, excluding the nucleus.

Cytoskeleton A group of non-membrane-bound organelles that supports the cell. Some serve as conduits for the transport of various cell components.

Differentiation The series of biochemical and structural changes that groups of cells undergo in order to form a specialized tissue.

DNA (deoxyribonucleic acid) The substance of heredity; a large molecule that carries the genetic information necessary for all cellular functions, including the building of proteins. DNA is composed of the sugar deoxyribose, phosphate, and the bases adenine, thymine, guanine, and cytosine.

DNA Structure DNA has a regular repeating chemical structure, called a double helix. There are, however, several forms of this helix: A, a form of right-handed DNA; B, the most common and most stable form of right-handed DNA; and Z, so-called left-handed DNA.

Domain A region (for instance, of a protein) characterized by a specific structural feature or features.

***Eco*RI** A restriction enzyme. Such enzymes allow researchers to cut DNA at specific sequences and thus help make recombinant DNA technology possible.

Electron Microscope A powerful microscope that uses beams of high-speed electrons instead of light waves to illuminate objects for observation.

Endocytosis The uptake of large molecules by the cell membrane. During this process, bits of the membrane fold inward and eventually pinch off to form small vesicles that move into the cytoplasm.

Endoplasmic Reticulum An organelle made up of membranes that form a system of tubes and flattened sacs continuous with the nuclear membrane. Some of the membranes are smooth (the SER); others are "rough" (the RER) because they are dotted with ribosomes.

Enzyme A substance (usually a protein) that speeds up, or catalyzes, a chemical reaction without being permanently alerted or consumed.

Eukaryotic Cell A cell that has a true nucleus surrounded by a membrane. This group includes all animal and plant cells, except blue-green algae.

Exocytosis The movement of substances that are packaged in vesicles to the cell surface, where they fuse with the cell membrane and release their contents outside the cell.

Exon A DNA sequence that codes for all or part of a gene product. In mammalian DNA, exons are separated by intervening sequences, called introns, that are copied into RNA and later removed as the RNA is processed.

Fluid-Mosaic Model A model of the cell surface membrane in which proteins move about within a bed of semi-fluid lipids.

Folding Problem The problem of why and how individual proteins fold into secondary, tertiary, and sometimes quaternary structures. See *protein structure.*

G Protein One of a group of proteins involved in the second messenger system that bind to phosphate in order to pass along an incoming signal.

Gene A unit of heredity; a segment of the DNA molecule containing the code for a specific protein product or function.

Genetic Code The language in which DNA's instructions are written; it consists or triplets of nucleotides, with each triplet corresponding to one amino acid in a protein structure or to a signal to start or stop protein production.

Glycolipid A molecule composed of sugar and fat that forms an important component of cell membranes.

Golgi Apparatus An organelle composed of membranous sacs that packages proteins into vesicles and sends them to the cell's surface or to lysosomes.

Heavy Metal A metal atom with a large number of protons, neutrons, and electrons; useful as an atomic marker in X-ray crystallography because it scatters X-ray beams strongly.

Helix The three-dimensional curve of a line in the shape of a coil, like the curve of a screw.

Heme Group The oxygen-carrying component of the hemoglobin and myoglobin molecules that also furnishes the deep-red pigment.

Hemoglobin The protein in red blood cells that transports oxygen through the bloodstream.

Hydrogen Bond A weak bond formed by the sharing of a hydrogen atom by two atomic nuclei.

Hydrophilic Having a strong affinity for water.

Hydrophobic Seeking to avoid water.

Intermediate Filament A component of the cytoskeleton that acts to strengthen the cell.

Intron A DNA sequence that interrupts the sequences coding for a gene product (exons). After information from the genes is transcribed into new strands of RNA, the introns are cut out of the RNA. The function of introns is still being explored.

Ion Any atom or small molecule that contains an unequal number of electrons and protons and, therefore, carries a net positive or negative electrical charge.

Light Microscope An instrument that magnifies objects using curved lenses and white light as a source of illumination.

Lipids Fats and fat-like compounds.

Liposome An artificial bubble made up of lipids that can contain substances, including drugs, designed to be absorbed by specific cells.

Lysosome A small organelle containing powerful enzymes that can digest a variety of materials.

Messenger RNA The ribonucleic acid molecule that transmits the genetic molecule that transmits the genetic information from the nucleus of a cell to the cytoplasm, where it directs protein synthesis.

Microfilament A threadlike organelle involved in cell motion, particularly muscle contraction.

Micrometer (or micron) One one-thousandth of a millimeter; 10,000 angstroms; convenient for describing the dimensions of cells and organelles.

Microtubule A thin, tubular organelle that acts as a structural support for the cell. During cell division, microtubules form the spindle that directs chromosomes to the daughter cells.

Mitochondrion The cell organelle that converts the energy in sugars into ATP, thereby fueling the cell.

Molecular Dynamics The study – usually through calculations requiring a high-speed computer – of the movement of atoms inside a protein or nucleic acid molecule and how this movement influences the protein's activity.

Molecule The smallest physical unit of an element or compound. A molecule of an element consists of one or more identical atoms. A molecule of a compound consists of two or more different atoms.

Mononucleotide-Binding Fold A recurring pattern in many proteins that bind to nucleotides, consisting of two beta strands connected by a crossover alpha helix. This fold has been found in proteins that bind to nucleotides even when the amino acid sequences of these proteins appear totally unrelated.

Mutation A change in the composition of DNA that is often reflected in a change in the protein for which it codes.

Myoglobin A protein that takes oxygen from red cells in the bloodstream and stores it in muscle tissue until needed.

Nanometer One one-thousandth of a micrometer.

Newton's Law of Motion Three classical laws of mechanics. First: a body remains at rest or in motion at a constant velocity unless acted upon by an outside force. Second: the sum of the forces acting on a body equals the body's mass times its acceleration. Third: for every force action on a body, the body exerts an equal and opposite force along the same line of action as the original force.

Nuclear Magnetic Resonance Spectroscopy (Abbreviated NMR) This technique measures the absorption of radio waves by atomic nuclei, most commonly hydrogen nuclei, placed in a magnetic field. NMR gives information about the environment of these nuclei and the distances between neighboring atoms which, when analyzed by computer, helps to determine the molecule's three-dimensional structure.

Nucleic Acid Either of two kinds of molecules (DNA and RNA), formed by chains of nucleotides, that carry genetic information.

Nucleosome The structural unit of DNA and protein in the nucleus. The structure of nucleosomes may hold answers to questions about how supercoiled DNA unwinds enough to allow genetic processes to occur.

Nucleotide A subunit of DNA or RNA. It includes one base, one phosphate molecule, and one sugar molecule (deoxyribose in DNA, ribose in RNA). See *base*.

Nucleus A membrane-bound subcellular structure that contains the genetic material.

Organelle A specialized structure having a definite function in a cell; for example, the nucleus, a mitochondrion, a ribosome.

Peptide A short string of amino acids joined by peptide bonds.

Peptide Bond The chemical bond that holds the amino acid subunits of a protein together.

Peroxisome A membrane-bound organelle that both generates and breaks down hydrogen peroxide.

Phospholipid A fatty compound that contains phosphate. Phospholipids make up much of the outer membranes of cells and organelles.

Polymer A compound composed of many repeating subunits linked in a chain.

Polypeptide A chain of amino acids.

Prokaryotic Cell A cell that does not have a membrane around its nuclear region; for example, a bacterium.

Protein A molecule made up of a number of amino acids arranged in a specific order determined by the genetic code. Proteins are essential for all life processes.

Protein Structure (primary, secondary, tertiary, quaternary) The primary structure is the order of amino acids in a linear chain. Immediately after its creation, the chain forms helices, sheets, and other shapes that make up the secondary structure. These units then interact to cause further folds and coils that make up the three-dimensional, or tertiary, structure. In some cases, several separate polypeptide chains may bind together into a single large protein molecule, forming the quaternary structure.

Receptor A specialized part of a cell that receives selected information about the environment and that may help to convey this information to other parts of the cell. The information is transmitted in the form of a specific chemical that must fit the receptor like a key in a lock.

Recombinant DNA Technology Techniques for cutting apart and splicing together pieces of DNA from different sources. When segments of human or animal DNA are transferred into another cell line or organism, these segments replicate and may be expressed together with the native genetic material of the cell line or organism.

Regulatory Protein A protein that controls gene expression by binding to the genetic material.

Replication The duplication of hereditary material prior to cell division.

Repressor A protein that shuts off gene expression.

Respiration Within cells, the breakdown of food molecules to liberate metabolically useful energy.

Restriction Enzyme An enzyme that recognizes a specific nucleotide base sequence (usually four to six base pairs in length) in a double-stranded DNA molecule and that cuts both strands of the DNA molecule at every place where this sequence occurs.

Rhinovirus A class of viruses that cause colds.

Ribosome An organelle that contains RNA and protein, and is the site of protein synthesis.

RNA (ribonucleic acid) A single-stranded nucleic acid that contains the sugar ribose. There are many forms of RNA, including messenger RNA, transfer RNA, and ribosomal RNA (all involved in protein synthesis), as well as several small RNAs whose functions are unclear. Certain viruses have RNA, instead of DNA, as their genetic material.

Second Messenger System A multi-step signal amplification process used by the cell to transmit, for example, signals from many hormones that cannot enter the cell directly.

Sequence The order of bases in a nucleic acid or of amino acids in a protein molecule.

Side Chain A chain of atoms attached to the central carbon atom of an amino acid. The side chain is the feature that distinguishes one amino acid from another.

Site-Directed Mutagenesis A recombinant DNA technique that permits scientists to replace, delete, or insert single amino acids in a protein. When the resulting mutant proteins are studied and compared to the natural protein, much can be learned about how amino acid changes affect the protein's function.

Structural Biology The field of science that seeks to explain the activity of biologically important molecules, such as nucleic acids and proteins, in terms of their atomic structure.

Structural Motif A specific type of protein structure formed by a combination of elements of secondary structure; e.g., a helix-turn-helix motif.

Synchrotron A massive electron storage ring with which physicists probe the fundamental components of matter. Since the X-ray beams of synchrotrons are thousands of times more intense than conventional X-ray beams, crystallographic data can be collected in a fraction of the usual time.

Tight Turn A common feature of protein structures that generally occurs at the surface of a protein. Its role is to interrupt or join together structures that are more internal, and it exerts a strong influence on the process of protein folding. Tight turns and "beta bulges' (irregularities in certain beta sheets) often indicate the presence of a protein's active site.

Transcription The transfer of information from various parts of the DNA molecule to new strands of messenger RNA, which then carry this information from the nucleus to the cytoplasm.

Translation The conversion of the genetic instructions for a protein from nucleotides of messenger RNA into amino acids.

Two-Dimensional NMR A variation of nuclear magnetic resonance spectroscopy that provides greater resolution of data, allowing these data to be used to study the structure and dynamics of small molecules.

Viral Protein Coat The protein cover that surrounds the genetic material of a virus and that is shed when the virus enters the cell.

Virus An infectious agent that is composed of a protein coat around a nucleic acid core. Viruses depend on living cells for their reproduction.

X-Ray Crystallography A technique for studying the three-dimensional structure of molecules by mathematically analyzing the patterns in which X-ray beams are diffracted by atoms in a crystal.

X-Ray Diffraction Alteration of an X-ray beam by the arrangement of atoms in the material through which it passes.

Zinc Fingers A class of regulatory proteins that bind to zinc or other metals as well as to DNA and RNA. They appear to be involved in the action of several hormone receptors that play important roles in disease.